Introduction to Climatology
For the Tropics

Introduction to Climatology For the Tropics

J. O. AYOADE

Department of Geography
University of Ibadan

JOHN WILEY & SONS

Chichester · New York · Brisbane · Toronto · Singapore

Library of Congress Cataloging in Publication Data:

Ayoade, J. O.
Introduction to climatology for the tropics.
Includes index.
1. Tropics—Climate. I. Title.
QC993.5.A96 551.6913 82-2648
ISBN 0 471 10349 7 (cloth) AACR2
ISBN 0 471 10407 8 (paper)

British Library Cataloguing in Publication Data:

Ayoade, J. O.
Introduction to climatology for the tropics.
1. Climatology
I. Title
551.6 QC981
ISBN 0 471 10349 7 (cloth)
ISBN 0 471 10407 8 (paper)

Photosetting by Thomson Press (India) Limited, New Delhi and printed at Vail-Ballou Press Inc., New York.

Contents

List of Figures

List of Tables

Preface

This book is aimed primarily at students and teachers of geography and related environmental sciences in universities and colleges in the tropics. The book is a basic text on the fundamental principles of climatology written with the needs of the readers in the tropics in mind. These needs have generally been glossed over by existing tests written by authors in the mid-latitudes primarily for readers in the same parts of the world.

In this book, atmospheric processes and weather systems are described and explained in a non-mathematical way since many of the students for whom the book is intended have only a limited knowledge of mathematics and physics. Throughout the book, emphasis is on both the physical processes in the atmosphere and their significance to man and his activities. The interactions between atmospheric processes and man are examined in some detail in chapters 12 and 13. These chapters provide a good groundwork material for students wishing to specialize in applied climatology. Agriculturalists, architects and planners working in the tropics will also find these and other chapters in the book useful and relevant to their needs.

The book is well referenced and illustrated with an appendix of revision questions on chapter by chapter basis to guide students in their reading.

I wish to express my gratitude to the Heads of Geography Departments at the Universities of Ibadan and Birmingham for providing needed technical and secretarial facilities during the preparation of this book. Thanks are also due to two anonymous readers for critically reviewing the manuscript and offering suggestions. Any errors or deficiencies that remain are, of course, my own responsibility.

J. O. AYOADE

CHAPTER 1

Introduction

The study of weather and climate occupies a central and important position within the broad field of environmental science. Atmospheric processes influence processes in the other parts of the environment, namely the biosphere, the hydrosphere and the lithosphere. Similarly, processes in these other parts of the environment cannot be ignored by the student of weather and climate. The four global realms—the atmosphere, hydrosphere, lithosphere and biosphere—are not only superimposed on one another but continually exchange matter and energy between them. As shown in Fig. 1.1 climate influences plants, animals (including man) and soil directly. It influences rocks through weathering while the external forces shaping the earth's surface are primarily controlled by climatic conditions. On the other hand, climate, particularly near the ground, is influenced by landscape features, vegetation and man through his various activities. Geomorphological, pedological and ecological processes and the

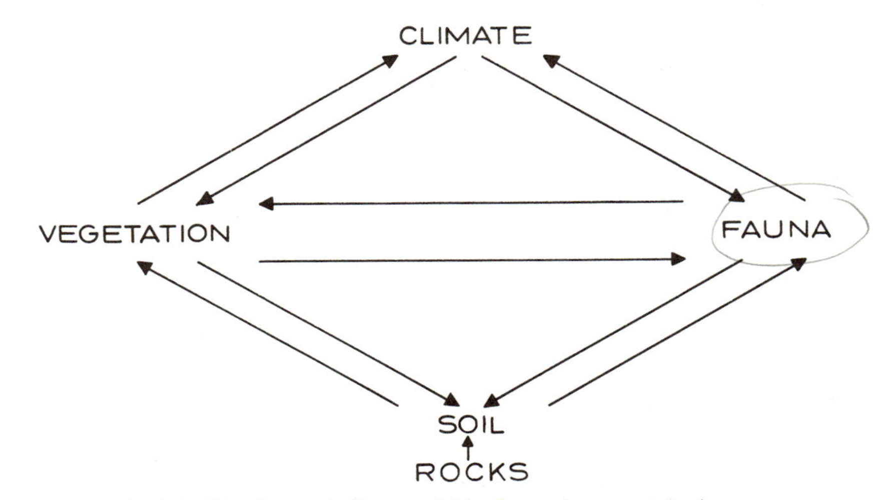

Fig. 1.1. Weather and climate within the environmental sciences

forms they give rise to can only be properly understood with reference to the climate prevailing now or in the past.

On weather and climate

In atmospheric science, a distinction is usually made between *weather* and *climate* and between *meteorology* and *climatology*. By weather we mean the state of the atmosphere at a given point in time at a given location. On the other hand, climate is the synthesis of weather at a given location over a period of about 30–35 years. Climate therefore refers to the characteristic condition of the atmosphere deduced from repeated observations over a long period. Climate includes more than the average weather conditions over a given area. It includes considerations of departures from averages (i.e. variabilities), extreme conditions, and the probabilities of frequencies of occurrences of given weather conditions. Thus, climate represents a generalization whereas weather deals with specific events.

Meteorology is broadly defined as the science of the atmosphere and is concerned with the physical, dynamical, and chemical state of the atmosphere and the interactions between the earth's atmosphere and the underlying surface. Climatology is the scientific study of climate as defined earlier. There is as would be expected, a considerable overlap in the contents of meteorology and climatology. The meteorologist and the climatologist, however, differ significantly in their methodology. Whereas the meteorologist employs the laws of classical physics and mathematical techniques in his study of atmospheric processes, the climatologist relies mainly on statistical techniques in deriving information about climate from weather information. It may be said, therefore, that the meteorologist studies the weather while the climatologist studies climate. However, climatology is founded on meteorology which is itself founded on the principles of physics and mathematics. There is therefore a close relationship between meteorology and climatology. Meteorology embraces both weather and climate while elements of meteorology must of necessity be incorporated in climatology to make the latter meaningful and scientific. Weather and climate can together be regarded as a consequence of, and a demonstration of the workings of the complex processes in the atmosphere, the oceans, and the land.

Nature and scope of climatology

As mentioned earlier, climatology deals with the patterns of behaviour of the atmosphere experienced over a long period of time. It is concerned more with the results of processes at work in the atmosphere than with their instantaneous operations. The scope of climatology is rather wide and subdivisions of the subject can be made on the basis of the topics emphasized or the scale of atmospheric phenomena that are emphasized. Under topical subdivisions of climatology we have the following, among others:

1 *regional climatology*—the description of climates over selected areas of the earth.
2 *synoptic climatology*—the study of the weather and climate over an area in relation to the pattern of prevailing atmospheric circulation. Synoptic climatology is thus essentially a new approach to regional climatology.
3 *physical climatology*—involves investigating the behaviour of weather elements or processes in the atmosphere in terms of physical principles. Emphasis is on the global energy and water balance regimes of the earth and the atmosphere.
4 *dynamic climatology*—emphasis is on atmospheric motions on various scales, particularly the general circulation of the atmosphere.
5 *applied climatology*—the application of climatological knowledge and principles to the solution of practical problems facing mankind.
6 *historical climatology*—the study of the development of climates through time.

Several other subdivisions are recognized in the literature. These include, for instance, agricultural climatology, bioclimatology, building climatology, urban climatology, statistical climatology, etc. These subdivisions can, however, be subsumed under one or more of the six subdivisions recognized above. Agricultural climatology, bioclimatology, and building climatology are, for example, aspects of applied climatology.

An alternative approach to the subdivision of climatology is based on scales of meteorological motion systems (Table 1.1). It must be emphasized, however, that the various atmospheric phenomena ranging from planetary waves to local wind systems constitute a single continuous spectrum of weather systems. Using the scalar system in Table 1.1, the following three subdivisions of climatology can be recognized:

1 *macroclimatology*—concerned with features of the climates of large areas of the earth and the large scale atmospheric motions that cause the climate.
2 *mesoclimatology*—concerned with the study of climate over relatively small areas of between 10 and 100 km across (e.g. study of urban climate or severe local weather systems like tornadoes and thunderstorms).
3 *microclimatology*—concerned with study of climate close to the ground surface or of very small areas less than 100 metres across.

Table 1.1 Scales of meteorological motion systems (after Barrett, 1974)

Motion system	Horizontal scale (km)	Vertical scale (km)	Time scale (hours)
Macroscale			
1 Planatery waves	5×10^3	10	2×10^2 to 4×10^2
2 Synoptic perturbations	5×10^2 to 2×10^3	10	10^{-2}
Mesoscale phenomena	$1–10^2$	1–10	1–10
Microscale phenomena	Less than 10^{-1}	Less than 10^{-2}	$10^{-2}–10^{-1}$

The development of modern climatology

Man has for a long time been interested in weather and the study of weather is as ancient as man's curiosity about his environment. This is hardly surprising since weather influences man and his diverse activities in numerous ways. The air man breathes, the food he eats and the water he drinks are all weather related. Even the way he lives—particularly his clothing and form of shelter—is to a large extent determined by weather. At first, man's understanding of weather phenomena was very poor. Weather phenomena were thought to be controlled by the gods until about the fifth century BC when the Greeks began to make meteorological observations. This brought a new and more scientific attitude to weather study as exemplified in *Airs, Waters and Places* written by Hippocrates about 400 BC and *Meteorologica* written by Aristotle some fifty years later.

But the rapid development of the science of the atmosphere had to wait for the technological revolution of the Renaissance period. In 1593 Galileo designed the thermometer and in 1643 the principle of the mercurial barometer was discovered by Torricelli, one of Galileo's pupils. The year 1832 saw the invention of telegraphy and weather data could then be gathered from a large number of widely located stations within minutes of the observations being made. Since then further technical developments in instrumentation, weather observations, and transmission and analysis of meteorological data have played vital roles in the development of modern meteorology and climatology.

The way the atmosphere is studied has also changed through time and particularly over the last thirty years or so. Traditional climatology is primarily concerned with describing the spatial and temporal distribution patterns of the weather elements over areas ranging in size from 1 or 2 km^2 to the whole earth. The method of description is cartographic consisting mainly of mean maps or graphs showing diurnal and seasonal changes and spatial differences in the values of the climatic elements such as temperature, precipitation, pressure, humidity, wind speed and direction, cloud amount etc. Climatic classification is also carried out usually in terms of the distribution of the climatic elements mentioned above.

This essentially descriptive approach to the study of weather and climate has several deficiencies and has led to misconceptions as to how atmospheric processes operate. Four of these deficiencies have been identified and discussed by Atkinson (1972). The first criticism of traditional climatology is that it is descriptive rather than explanatory. The mean maps of elements are essentially descriptive and do not give any idea about the processes causing this distribution. Second, the traditional approach to the study of weather and climate tends to give the impression of a static atmosphere whereas the atmosphere is dynamic and in constant turmoil. The atmospheric characteristics at a given place can change on time scales ranging from microseconds to hundreds of years. The use of 30–35 years for averaging values of climatic elements under traditional climatology does not take into consideration the continuous changes taking place within the atmosphere. Third, the traditional method of studying weather and climate tends to neglect interactions, i.e. feedback mechanisms operating in the atmosphere. Processes interact and affect one another and effects often rebound to change or modify their causes. As Atkinson (1972) has pointed out, such feedback mechanisms are vital to the constant struggle of the atmosphere to iron out extremes and achieve an elusive state of equilibrium. The fourth criticism of methods of traditional climatology is in respect of its climatic classification. Boundary lines on climatic maps give the erroneous impression that climate changes abruptly at the lines; of course this does not happen. What we have is a gradation in characteristics from one climatic type into another. Also, the climates of the defined areas are often thought of as separate climatic entities and explained as such usually by reference to surface phenomena only. This approach is wrong as it ignores the fact that climate has a third dimension in the vertical and that the atmospheric characteristics at a given place can only be meaningfully explained when viewed in the context of the workings of the whole atmosphere. As Atkinson (1972) has emphasized, trying to explain local atmospheric circulations solely in terms of local factors is like trying to divide an indivisible whole.

Modern climatology aims at removing the deficiencies of traditional climatology outlined above. Emphasis is now given to explaining atmospheric phenomena in addition to describing them. The atmosphere is dynamic, not static, and efforts are now being made to understand the processes and interactions taking place within the atmosphere and at the atmosphere–earth interface.

The emergence of modern climatology can be ascribed to two factors, namely the challenges posed by needs of the society and improvement in data collection and analysis as discussed below. Traditional climatology with its emphasis on description is of little practical usefulness to man. Modern man is affected by weather just like his primitive ancestors. But unlike his forebears, modern man does not want to live at the mercy of weather. He therefore wants to manage or even control weather. To do this, man must be able to understand weather phenomena so that he can predict, modify or control them where possible. Hence, there is a need for emphasis on the explanation of weather processes which, as mentioned earlier, is the hallmark of modern meteorology. To predict or forecast weather we have to understand the workings of the atmosphere. The challenges posed to meteorologists and climatologists have been mainly in agriculture and aviation. Challenges have also arisen from the need to protect man and his property against the effects of extreme weather events. The value of weather knowledge in commerce and industry is also now generally acknowledged.

Meteorological observations have come a long way since the early development of the wind vane and the rain gauge in the fifth century BC. Observations are now taken from various platforms ranging from the earth's surface, through balloons, helicopters, and aircrafts to rockets and satellites. There have been significant improvements in the design and accuracy of the meteorological instruments to be found in conventional weather stations. Increasing use is now being made of *radiosonde* to sample upper atmospheric characteristics.

Similarly, *radar* is now widely being used for the routine forecasting of weather and for research into cloud physics and dynamics (see Chapter 9).

A major breakthrough in the observation of weather has been the development of meteorological satellites. Weather satellites give us objective, large area coverage of weather systems and enable us to measure radiation from a position outside the earth's atmosphere. As Barret (1974) has pointed out, modern meteorological satellites are performing three important roles as

1 observing systems of the earth and its atmosphere,
2 data collection platforms, and
3 communication links between widely spaced ground stations which have to exchange weather data daily.

Weather satellites now constitute important sources of climatological information for the atmospheric scientist. Weather satellites have greatly improved the data coverage of the earth as they now provide information on weather in remote, inhospitable, or uninhabited areas of the world particularly the oceans and the deserts as well as the tropics and polar areas which are not well served with conventional meteorological stations. Meteorological satellites have other advantages. Satellite-derived meteorological data are internally more homogeneous than those obtained from the conventional weather stations. Satellite-derived data are also spatially continuous across the surface of the earth unlike most data from conventional weather stations which are point measurements of the atmosphere from sample sites. Besides, satellites can provide a more frequent data coverage than the conventional weather stations which usually report every six, twelve, or twenty-four hours. Satellite-derived data also lend themselves readily to computer processing.

There are, however, problems associated with satellite operation and the analysis and usage of meteorological data obtained from them. It must be borne in mind that satellite observations are really complementary in nature to conventional measurements and cannot replace them. This is because satellites are remote sensing platforms investigating the atmosphere by instruments not in direct contact with it. Their observations are therefore different in nature from those made within the atmosphere itself by instruments in direct contact with the system.

Satellite-derived data are already very voluminous and there are problems of data selection and reduction as well as processing, analysis, and interpretation. The problem is compounded by the fact that the resolutions of the data may not be optimal for climatological purposes. Also, the quality of data from satellites tends to deteriorate with time because of degradation in satellite sensory systems with the passage of time. Partly because of this, data from two or more satellites are often not comparable.

These problems are, however, not insurmountable and efforts are already being made to solve them. The electronic computer has been of much help in storing, processing, and analysing data derived from satellites. Apart from the uses of satellite-derived data in weather forecasting and modelling of the general circulation of the atmosphere, climatological uses to which weather satellite data are being put include the following, among others (see Barret 1974, 1975):

1 completing maps of climatic elements such as cloud cover particularly in areas with sparse conventional data coverage.
2 mapping climatic variables which formerly could not be measured or were not considered. A good example is the pattern of net radiation balance at the top of the earth's atmosphere or radiation temperatures of cloud tops exposed to space.
3 classifying climates on new bases using net radiation, vorticity distributions and humidity regimes.
4 estimating rainfall in areas without adequate raingauge network.

Recent developments in tropical climatology

The tropics has been variously defined as:

1 the area between the Tropics of Cancer and Capricorn which indicate the outer limits of the areas where the sun can ever be in zenith.
2 the area between latitudes 30° north and south of the equator.
3 the area of the world where there is no cold season, i.e. where winter never comes.
4 the area of the world where the annual range of temperature is equal to or less than the mean daily range.
5 the area of the world where the mean sea level temperature for the coldest month of the year is not below 18 °C.
6 that part of the world where the weather sequences differ distinctly from those of middle latitudes, the dividing line between easterlies and westerlies in the middle troposphere serving as a rough guide to the boundary (Riehl, 1954).

It is clear from the foregoing that the area of the world referred to as the tropics is characterized by the absence of a cold season and considerable diurnal range of temperatures. Seasons are recognized within the tropics primarily on the

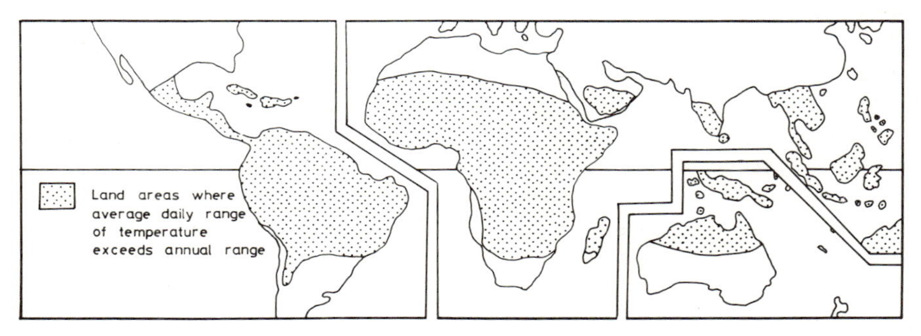

Fig. 1.2. The extent of the tropics

basis of rainfall occurrence and the air's relative humidity. On the basis of the annual rainfall totals received, the tropics can be divided into the humid tropics where the mean annual rainfall is more than 600 mm and the dry tropics where it is less than 600 mm (see Fig. 1.2). The tropics as delimited in Fig. 1.2 constitute 40% of the land surface of the earth and are inhabited by some 1400 million people or 40% of the world population (Nieuwolt, 1977). Most tropical countries are poor, agricultural countries with low per capita income. The tropics is therefore an area in need of development. Climate is a vital component of the tropical environment which must be understood and taken into account in any development programme aimed at raising the standard of living and the quality of life of the 1400 million inhabitants.

Compared to the temperate lands and even the polar zones, the tropical lands have suffered from poor data coverage and the development of indigenous models and analytical techniques for analysing and studying the prevailing weather and climate. Although as mentioned earlier the data situation in the tropics has been greatly improved with the advent of the weather satellites the tropics are still poorly served by conventional weather stations. Upper air stations are still grossly inadequate. The tropics are mostly inhabited by people at a low state of economic development. Governments of tropical countries have few resources to allocate to the development of observation network in the face of competing demands from other sectors of the economy yearning for improvement. Because of the low literacy level, skilled personnel are in short supply in all sectors of the economy. Many areas are also remote or inhospitable and therefore unsuitable for the establishment of reliable and functioning weather stations. The tropical climatic conditions also affect the durability and accuracy of some meteorological instruments particularly the more sophisticated and sensitive ones.

Thus, lack of requisite data has influenced the course of research and development of meteorology and climatology in the low latitudes. Intensive study of weather and climate in the tropics dates only from World War II when many weather stations were established in these areas for aviation purposes. At least three descriptive models of tropical climate have been put forward. These are the models by:

1 the climatological school,
2 the air mass school, and
3 the perturbation school.

The model by the climatological school considers tropical weather in terms of diurnal and seasonal controls plus local effects. According to this model, day-to-day weather in the tropics differs very little from what is revealed by the monthly and annual means. Concepts like the trades, the monsoons, and the doldrums are used to explain patterns of tropical weather and climate.

The air mass–frontal model is an attempt to transfer the concepts of air mass and fronts developed by Bjerkness and Bergeron in the middle latitudes to the tropics with little or no modification. The model has proved to be misleading and useless in the tropics mainly because tropical air masses differ very little in their thermal characteristics.

The perturbation model of tropical weather and climate is to some extent founded on the concepts of the climatological model like the trades, monsoons, and the doldrums. But the perturbation school stresses the importance of perturbations to which the mean circulation patterns are subject, these perturbations being accompanied by characteristic weather and pressure patterns identifiable on the daily weather map.

After discussing the merits and demerits of the above three models or schools of thought regarding tropical weather and climate, Palmer (1951, p. 872) predicts that: 'future advances in tropical meteorology will come through the application of the concepts and methods of the perturbation school provided they are freed from fallacious generalisations'. Palmer has indeed been proved right! It is now generally agreed that we can talk about weather, not only climate, in the tropics. Besides, weather changes in the tropics are now known to be frequent and complex with quite distinct types of weather. Our view of tropical weather systems has been radically revised by satellite-derived data coming in since the 1960s. The various categories of weather systems in the tropics distinguished according to their space and time scales are discussed in Chapter 6. It is now generally recognized that mid-latitude synoptic models are inapplicable in the tropics. Attempts are now being made thanks to the weather satellites and improved observational network to develop tropical synoptic models and appropriate methods of meteorological analysis. The most important of these new developments in weather observations and analyses in the tropics are discussed in Chapter 9.

The role of WMO

The need for improvement in the collection of meteorological data particularly in the tropics cannot be overemphasized. Progress in the development of new concepts and theories in meteorology depends to a large extent on a good weather observational network and the free and quick exchange of meteorological information amongst the nations of the world. These developments have been made possible by the establishment in 1873 of the International Meteorological Organization (IMO) which on March 23, 1950 became known as the World Meteorological Organization (WMO). In December 1951, WMO was recognized as a specialized agency of the United Nations. The declared purpose of the organization was three fold:

1. to facilitate world-wide cooperation in the establishment of meteorological station networks;
2. to promote the development of centres for meteorological services; and
3. to promote rapid exchange of weather information and the standardization and publication of weather observations.

Today, the programme of scientific and technical activities of WMO can be classified into four broad categories:

1 the World Weather Watch (WWW);
2 the WMO Research Programme;
3 the WMO programme on the interaction of man and his environment;
4 The WMO Technical Cooperation Programme.

The World Weather Watch Project is designed to improve world weather services. The main features of the project include the following:

1 a more accurate and comprehensive global weather observation system involving the use of satellites and other advanced and automatic devices;
2 the establishment of three world meteorological centres at Melbourne (Australia), Moscow (USSR), and Washington (USA) together with a number of regional centres for storing and processing meteorological data;
3 the establishment of a global telecommunication system for rapid transmission of meteorological data, analyses, forecasts, and warnings;
4 the training of meteorologists of all cadres.

The WMO Research Programme is known as the Global Atmospheric Research Programme (GARP) and is designed to develop and test fundamental physical and mathematical bases of long-range weather forecasting. GARP is composed of several auxiliary programmes known as GARP subprogrammes. Examples include GARP Atlantic Tropical Experiment (GATE) and West African Monsoon Experiment (WAMEX). The WMO programme on the interaction of man and his environment is aimed at the application of meteorological knowledge to human activities such as agriculture, transportation, and the use and development of water resources. Under the WMO Technical Cooperation Programme the WMO assists countries in the development of their meteorological services and the training of personnel.

It is quite clear from the foregoing that the WMO has played and will continue to play an important role in the development of the science of meteorology. All its programmes are laudable and they deserve to be supported morally and financially by all.

References

Atkinson, B. W. (1972). The atmosphere. In Bowen, D. Q. (ed.), *A Concise Physical Geography*. Hulton Educational Publications, London.
Barret, E. C. (1974). *Climatology from Satellites*. Methuen, London.
Barret, E. C. (1975). Analysis of image data from meteorological satellites. In Peel, R. *et al.*, (eds.), *Processes in Physical and Human Geography*. Heinemann, London.
Nieuwolt, S. (1977). *Tropical Climatology*. John Wiley, London.
Palmer, C. E. (1951). Tropical meteorology. In Malone, T. F. (ed.), *Compendium of Meteorology*. American Meteorological Society, Boston, U.S.A.
Riehl, H. (1954). *Tropical Meteorology*. McGraw-Hill, New York.

CHAPTER 2

The Earth's Atmosphere

The composition of the atmosphere

The atmosphere can be described as a thin layer of odourless, colourless, and tasteless gases held to the earth by force of gravity. The atmosphere comprises a stable mechanical mixture of gases, the most important of which are nitrogen, oxygen, argon, carbon dioxide, ozone, and water vapour (see Table 2.1). Other gases occur in very small proportions and they include neon, krypton, helium, methane, hydrogen, etc. Nitrogen, oxygen, and argon are constant in amount but carbon dioxide, ozone, and water vapour vary in amount both spatially and temporally. For instance, the amount of water vapour in the atmosphere can vary practically from zero in arid areas to about 3–4% in the humid tropics. The water vapour content of the atmosphere is closely related to air temperature and availability of water at the earth's surface. Thus, in the middle latitudes it is greater in summer than in winter when the capacity of the atmosphere to hold moisture is small. Water vapour is also almost absent at about 10–12 km above the earth's surface. This is because the water vapour in

Table 2.1 Average composition of the dry atmosphere below 25 km (after Barry and Chorley, 1976)

Gas	Volume % (dry air)
Nitrogen (N_2)	78.08
Oxygen (O_2)	20.94
Argon (Ar)	0.93
Carbon dioxide (CO_2)	0.03 (variable)
Neon (Ne)	0.0018
Helium (He)	0.0005
Ozone (O_3)	0.00006
Hydrogen (H)	0.00005
Krypton (Kr)	Trace
Xenon (Xe)	Trace
Methane (Me)	Trace

the atmosphere is supplied by evapotranspiration of water from the earth's surface and carried upwards by turbulence which is most effective below a height of 10 km.

Ozone is concentrated between the heights of 15 and 35 km within the atmosphere. The ozone content of the atmosphere is low over the equator and high over areas poleward of the 50° latitudes. Ozone is formed when under the influence of ultraviolet radiation oxygen molecules break up and the separated atoms individually combine with other oxygen molecules. Although the break-up of oxygen molecules usually takes place in the layer between 80 and 100 km, the formation of ozone takes place in the layer between 30 and 60 km. This is because of the very low density of the atmosphere at 80 and 100 km which does not encourage collisions between O and O_2, a process necessary for the formation of ozone. Ozone itself is unstable as it may be destroyed by the action of radiation on it or by collisions with monatomic oxygen (O) to recreate oxygen (O_2) as follows:

$$O_3 + O \rightarrow O_2 + O_2.$$

The distribution pattern of ozone within the atmosphere is thought to be the result of some circulation mechanism transporting ozone to suitable levels where its destruction is less likely and its concentration thus assured. Such areas are found in the atmosphere at elevations of 15–35 km above the earth's surface.

Carbon dioxide (CO_2) enters the atmosphere mainly by the action of living organisms in the ocean and on the land. Photosynthesis helps to maintain a balance in the amount of global carbon dioxide by removing about 3% of the world's total carbon dioxide annually. There are fears, however, that increasing use of fossil fuels by man is upsetting this balance and that there has been some increase in the amount of carbon dioxide in the atmosphere. For instance, the total quantity of carbon dioxide in the atmosphere between 1870 and 1970 has been estimated to have increased from 294 to 321 ppm (about 11% in increase) due to the burning of fossil fuels (Barry and Chorley, 1976).

In addition to the gases listed in Table 2.1 the atmosphere contains variable but significant quantities of aerosols. These are suspended particles of dust, smoke, organic matter, sea salt, etc. coming from both natural and man-made sources. Man-made aerosols are now estimated to account for about 30% of the total aerosols contained in the atmosphere and this figure may double by 2000 AD (Barry and Chorley, 1976).

Water vapour, ozone, carbon dioxide, and aerosols play important roles in energy distribution and exchanges within the atmosphere and between the earth's surface and the atmosphere. Their amounts and distribution patterns within the atmosphere should therefore be carefully studied. Rocket observations indicate that nitrogen, oxygen, and argon are mixed in constant proportions up to a height of 80 km owing to the constant stirring within the atmosphere. Contrary to what might be expected, there is no separation of light gases (e.g. hydrogen and helium) from the heavier ones within the atmosphere because of the constant large scale turbulent mixing of the atmosphere. However, as mentioned above, spatial and seasonal variations do occur in the distribution of aerosols, carbon dioxide, water vapour, and ozone. Because these aerosols and gases absorb, reflect, and scatter both solar and terrestrial radiation, the heat budget of the earth–atmosphere system and the temperature structure of the atmosphere are greatly affected by their amounts and distributions within the atmosphere.

The mass of the atmosphere

The atmosphere being a mechanical mixture of gases displays the principal characteristics of all gases. It is extremely mobile, compressible, and has capacity for expansion. These characteristics explain some of the fundamental features of atmospheric structure as well as many features of weather and climate. Because the atmosphere is highly compressible its lower layers are much more dense than those above. The average density of the atmosphere decreases from $1.2\,kg\,m^{-3}$ at the earth's surface to $0.7\,kg\,m^{-3}$ at a height of 5000 metres. In fact as shown in Fig. 2.1, about half the total mass of the atmosphere is found below 5 km. Atmospheric pressure decreases logarithmically with height above the earth's surface. The pressure at a point in the atmosphere is the weight of air vertically above a unit horizontal area centred on that point. In other words, it is the downward force on unit horizontal area resulting from the action of gravity on the mass of air vertically above (McIntosh and Thom, 1969). As we move up in the atmosphere, the air becomes thinner until the outer limit of

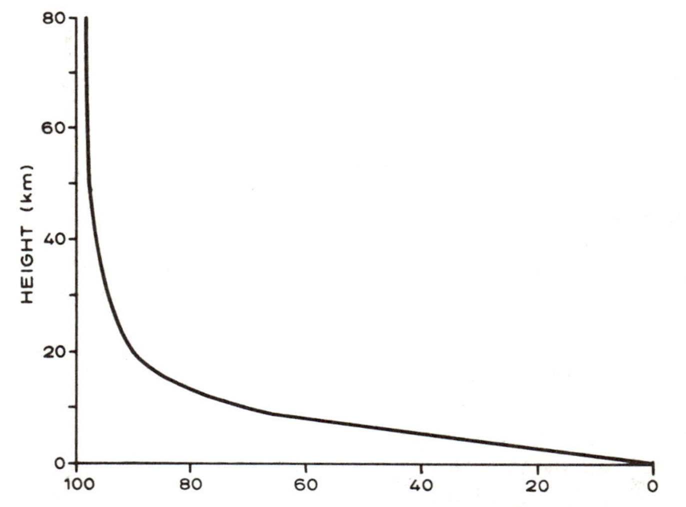

Fig. 2.1. Vertical distribution of the mass of the atmosphere (from Barry and Chorley, 1976). Per cent of total mass of the atmosphere lying below elevations upto 80 km. This shows the shallow character of the earth's atmosphere. Reproduced by permission of Methuen.

the atmosphere is reached and we move into the outer space. The actual density of air depends on the temperature, the amount of the water vapour in the air, and gravity. Since all these vary there is no simple relationship between altitude and pressure. Nevertheless, the relationship between pressure and altitude is so significant that meteorologists often express elevations in millibars, the unit of pressure measurement. For instance, 1000 mb represents sea level and 300 mb about 9000 metres.

Pressure is measured as force per unit area. The unit used by meteorologists is the millibar (mb). One millibar is equivalent to 1000 dynes per square centimetre. (A dyne is the force required to produce an acceleration of one cm/s^2 in a mass of one gram.) Pressure readings are made with mercury or aneroid barometer. These readings are influenced by variations in gravity over the earth's surface. To correct for these variations readings are referred to standard gravity values of 9.81 m s^{-2} and for 45° latitude. Mercury barometer readings also need to be standardized to a temperature of 0 °C to allow for the thermal expansion of mercury.

The average atmospheric pressure at sea level is 1013.25 mb. Each of the gases in the atmosphere exerts a partial pressure independent of the others. Thus, on average at sea level nitrogen exerts a pressure of 760 mb, oxygen 240 mb, and water vapour 10 mb. Pressure exerted by water vapour, i.e. vapour pressure, varies with latitude and season. For instance, it is about 0.2 mb over northern Siberia in January and over 30 mb in the tropics in July. But this variation is not reflected in the pattern of total surface pressure. In fact, owing to dynamic factors, air in high-pressure areas is generally dry while that in low-pressure areas is usually moist (Barry and Chorley, 1976).

The structure of the atmosphere

Evidence from rawinsonde, radiosonde, rockets, and satellites indicate that the atmosphere is structured into three relatively warm layers separated by two relatively cold layers (see Fig. 2.2). The three warm layers occur near the earth's surface, between 50 and 60 km and above 120 km while the cold layers are found between 10 and 30 km and at about 80 km above the earth's surface.

The lowest layer of the atmosphere is called the *troposphere*. The troposphere contains about 75% of the total gaseous mass of the atmosphere and virtually all the water vapour and aerosols. It is therefore the layer where weather phenomena and turbulence are most marked and has aptly been described as the weather-making layer of the atmosphere. For these reasons, it is of most direct importance to man. Within the troposphere temperature decreases with height at an average rate of 6.5 °C per kilometre. The top of the troposphere is called the tropopause and is characterized by temperature inversion conditions which effectively limit convection and other weather activities. The height of the tropopause is not constant. It varies from place to place and from time to time over a given area. The altitude of the tropopause is, however, highest at the equator (16 km) where there is intense heating and vertical convective

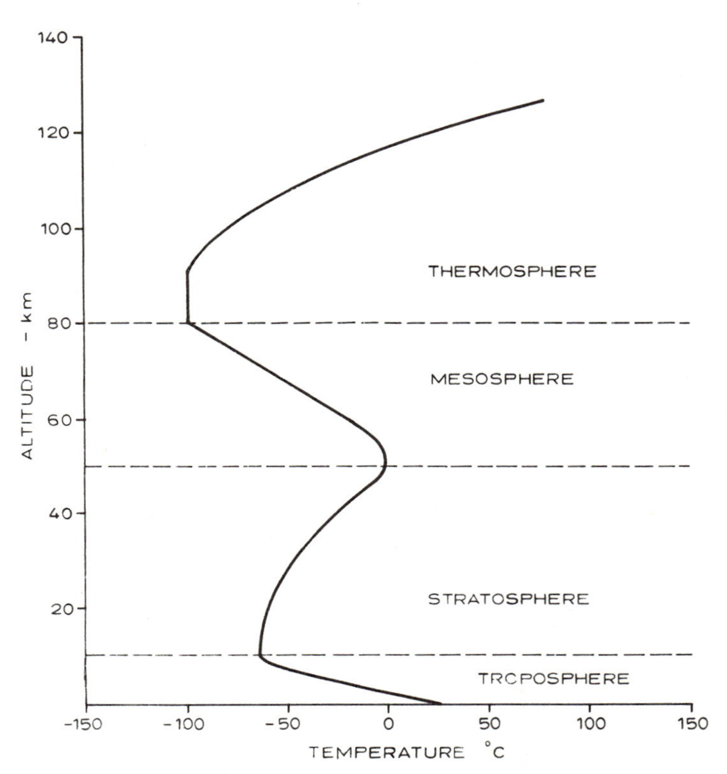

Fig. 2.2. The structure of the atmosphere according to temperature change

turbulence and least at the poles where it is only 8 km. The troposphere can itself be divided into three layers on the basis of the dominant mechanism for energy exchanges. These are the laminar layer, the frictional layer, and the so-called free atmosphere.

The laminar or surface layer marks the interface between the land and the atmosphere. Energy transfer within this layer is by conduction and vertical exchanges of heat and moisture are rather slow. On top of this layer is the friction layer which is about 1000 metres deep. Vertical heat transfer in this layer is mainly by turbulence or eddy motion. Next is the free atmosphere, that is the layer that is free from the effect of friction created by irregularities on the earth's surface. In this layer winds are stronger since they no longer suffer from frictional drag and vertical energy transfer is mainly through the formation of clouds. Water is evaporated from the earth's surface and carried upward as vapour. When the vapour condenses in the atmosphere to form clouds the latent heat is released.

The *stratosphere* is the second major layer of the atmosphere and extends from the tropopause to about 50 km above the ground. Unlike in the troposphere, temperature generally increases with height in the stratosphere. The air density is much less here so that even limited absorption of solar radiation by the atmospheric constituents, notably ozone, produces a large temperature increase. The stratosphere contains much of the total atmospheric ozone. Maximum ozone concentration occurs around the height of 22 km above the earth's surface. Unlike the troposphere, the stratosphere contains little or no water vapour. Marked seasonal changes of temperature are characteristic of the stratosphere and it is generally believed that events in the stratosphere are probably linked with temperature and circulation changes in the troposphere. The top of the stratosphere is marked by an isothermal zone called the stratopause. The troposphere and the stratosphere constitute the lower atmosphere.

The upper atmosphere is generally regarded as starting from the stratopause and terminating at where the atmosphere merges with the outer space. Several layers have been recognized within the upper atmosphere but as yet there is no universal agreement on terminology and the number of layers. The following layers are generally recognized.

1 The mesosphere, where temperature decreases with height to reach a minimum of about $-90\,^{\circ}C$ around 80 km, lies on top of the stratosphere. Atmospheric pressure is very low and decreases from about one mb, at the base of the mesosphere about 50 km above the ground to 0.01 mb at the mesopause about 90 km above the earth's surface.
2 The thermosphere is the next layer. Here, temperature increases with height owing to the absorption of ultraviolet radiation by atomic oxygen. The atmosphere is very tenuous as densities are extremely low. Above 100 km the atmosphere is greatly affected by solar X-rays and ultraviolet radiation which cause ionization or electrical charging. This region of high electron density is sometimes called the ionosphere.
3 The exosphere extends from a height of between 500 and 750 km above the earth's surface and beyond. Atoms of oxygen, hydrogen, and helium form a very tenuous atmosphere and the gas laws cease to be valid. The atmosphere has no real upper boundary but becomes progressively and rapidly less dense before it finally merges with the outer space.

It is clear from the foregoing that the earth's atmosphere varies in characteristics from the base upwards. As far as weather and climate are concerned, only the troposphere and the stratosphere, particularly the former, are of interest. The upper atmosphere is still relatively unexplored compared to the lower atmosphere.

References

Barry, R. G. and Chorley, R. J. (1976). *Atmosphere, Weather and Climate* (3rd eds.). Methuen, London.

McIntosh, D. H. and Thom, A. S. (1969). *Essentials of Meteorology*. Wykeham Publications, London.

CHAPTER 3

Radiation and the Heat Budget

Solar radiation

The sun provides 99.97% of the energy used for various purposes in the earth–atmosphere system. Every minute the sun radiates about 56×10^{26} calories of energy of which the earth intercepts only 2.55×10^{18} calories. Although this represents only one 2000 millionth of the total solar energy sent into space, it is estimated to be 30,000 times greater than the total annual energy consumption of the world.

The sun, a luminous gaseous sphere, has a surface temperature of 6000 °C and emits energy in electromagnetic waves which travel at the rate of about 299,300 km per second (see Fig. 3.1). The energy which travels radially outward from the sun takes only $9\frac{1}{3}$ minutes to travel about 150 million km, the distance between the earth and the sun. Although solar radiation travels through space without energy loss, the intensity of radiation decreases inversely as the square of the distances from the sun. The amount of solar energy received per unit area of surface held at right angles to the sun's rays at the top of the atmosphere is about 2 calories per cm^2 per minute or 2 langleys per minute. This is known as the *solar constant* because this amount is relatively constant, the variation around the mean value of 2 langleys per minute being approximately 2%.

The sun radiates like a black body. According to the Stefan–Boltzman law, the flux of radiation from a black body is directly proportional to the fourth power of its absolute temperature.

$$F = \sigma T^4 \tag{3.1}$$

where F is the radiation flux, T is absolute temperature of the body and σ is the Stefan–Boltzman constant. Black bodies also absorb all radiant energy falling on them. Most solids and liquids behave like black bodies but gases do not. At a given temperature the emission in each wavelength from a black body is the maximum possible. According to Wien's displacement law, the wavelength of maximum intensity of emission from a black body is inversely proportional to the absolute temperature of the body. Thus,

$$\lambda_{max}(\mu m) = 2897\, T^{-1}. \tag{3.2}$$

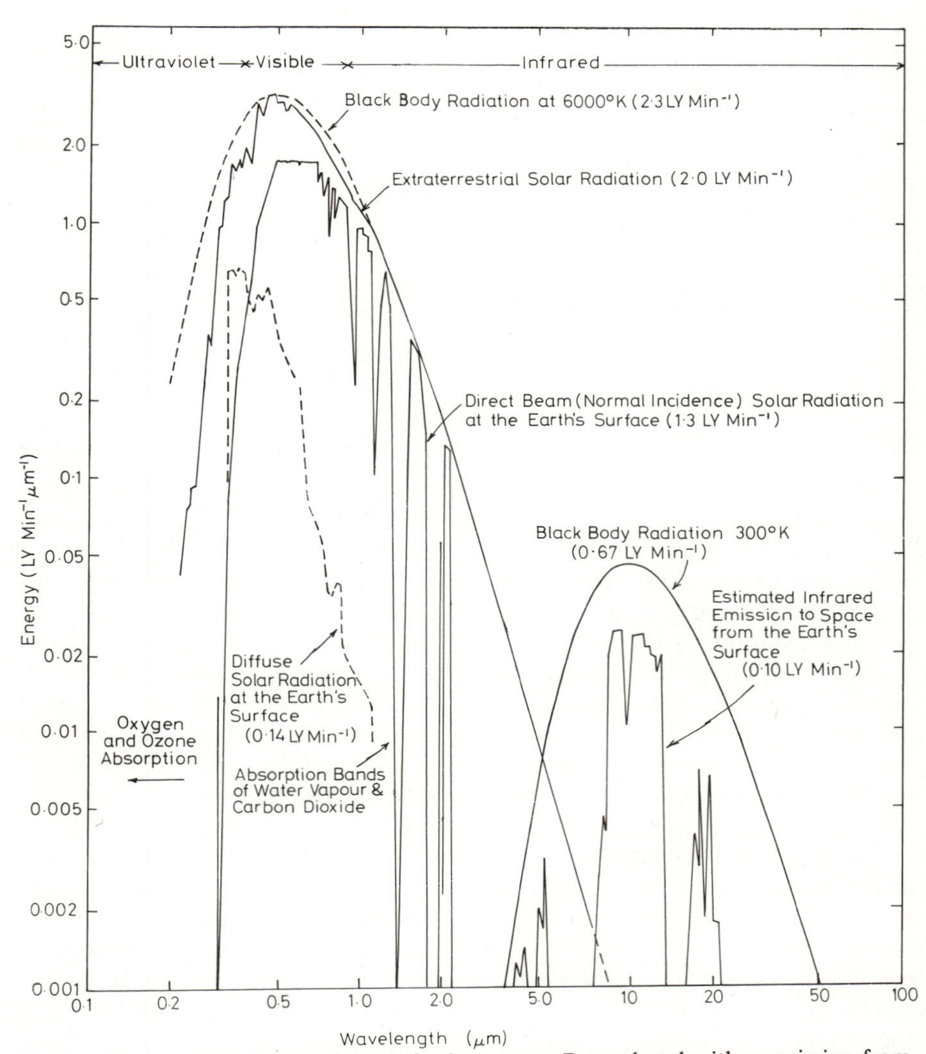

Fig. 3.1. Electromagnetic wavelength of solar energy. Reproduced with permission from Sellers, *Physical Climatology*, 1965. Published by The University of Chicago Press

For the sun the wavelength of maximum emission is near 0.5 micron (0.5 μ). Almost 99% of solar radiation is in the short wavelength from 0.15 to 4.0 μm. According to Sellers (1965) a breakdown of the spectral composition of solar radiation indicates that 9% is in the ultraviolet ($\lambda \leqslant 0.4$ μm), 45% is in the visible (0.4 μm $\leqslant \lambda \leqslant$ 0.74 μm), while the remaining 46% is in the infrared ($\lambda \geqslant 0.74$ μm).

The amount of solar radiation incident on the top of the earth's atmosphere depends on three factors namely the time of year, the time of day, and latitude. Fig. 3.2 shows the daily variation of solar radiation at the top of the atmosphere as a function of latitude. The distribution is not symmetric because the earth

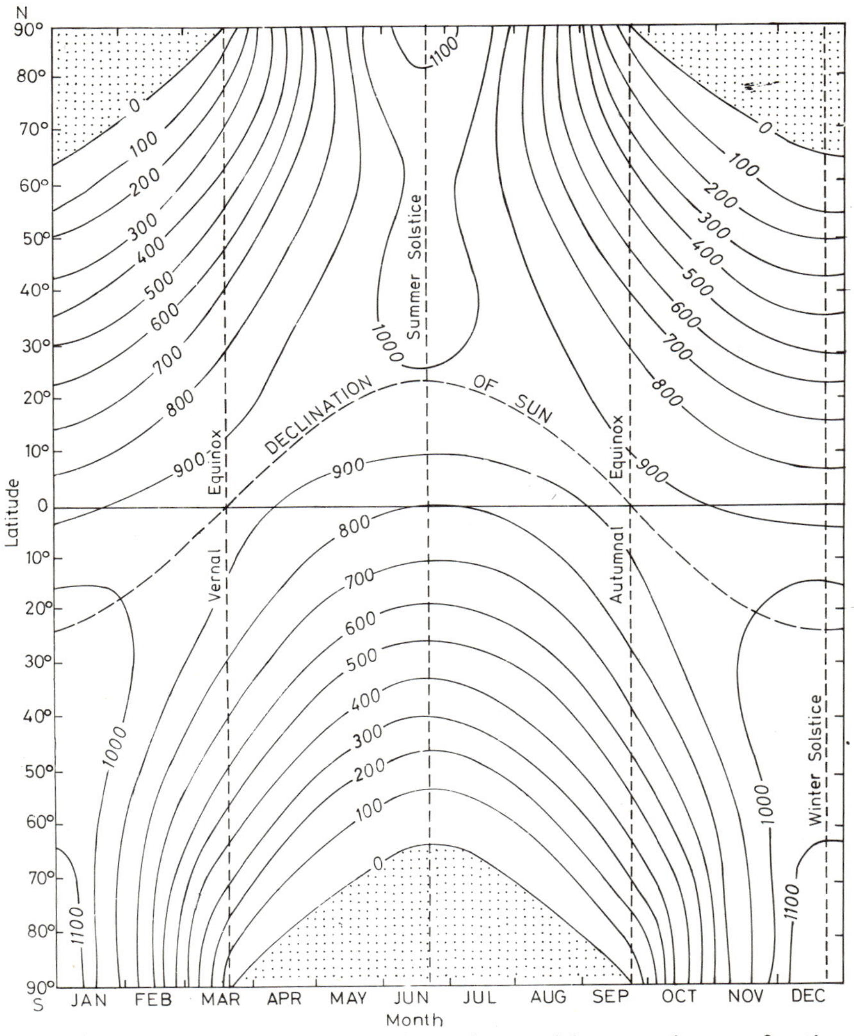

Fig. 3.2. Daily variation of solar radiation at the top of the atmosphere as a function of latitude in langleys per day. Reproduced with permission from Sellers, *Physical Climatology*, 1965. Published by The University of Chicago Press

is closest to the sun in January so that more radiation is received in all latitudes during the northern winter than during the northern summer (see Sellers, 1965). The distance of the earth from the sun varies through the year since the earth's orbit round the sun is elliptical rather than circular. These variations in distance affect the amount of solar energy received. For instance, the solar energy received by a surface normal to the beam is 7% more on 3 January at the perihelion than on 4 July at the aphelion.

The altitude of the sun, which is the angle between its rays and a tangent to the surface at the point of observation, also affects the amount of solar energy received. The greater the sun's altitude the more concentrated is the radiation intensity per unit area and the less the albedo (i.e. the proportion of incident radiation reflected by the surface). The sun's altitude is determined by latitude of the site, the time of day, and the season. The sun's altitude generally decreases with increasing latitude. It is high in the afternoon but low in the morning and evening. Similarly the sun's altitude is higher in summer than in winter.

The total amount of radiation received at a given location is also affected by the length of daylight. The duration of sunshine obviously affects the quantity of radiation received. The length of daylight varies with latitude and the season. Around the equator days and nights are almost of equal duration throughout the year. Day length generally increases or decreases with increasing latitude depending on the season. In summer for instance, day length increases from the equator towards the north pole and decreases towards the south pole. Between the Arctic circle and the north pole daylight lasts the entire 24 hours. During the winter solstice of the northern hemisphere the reverse is the situation. Day length increases towards the south pole but decreases towards the north pole. Also, between the Antarctic circle and the south pole daylight lasts the entire 24 hours while in similar latitudinal locations in the northern hemisphere the duration of night is 24 hours.

Finally, the amount of solar energy intercepted by the earth is obviously related to the total energy sent out into space by the sun (i.e. the solar output). As mentioned earlier, the solar output is not constant as evidenced by the slight variations of 1–2% in the value of the solar constant. These variations are probably related to the sunspot cycle but because the estimation of the solar constant is subject to errors of similar magnitude we cannot say with certainty whether or not there are fluctuations in the values of the solar constant.

The above factors namely distance from the sun, altitude of the sun, length of day, and solar output produce the pattern of receipt of solar energy at the top of the atmosphere shown in Fig. 3.2. The equator has two insolation maxima at the equinoxes and two minima at the solstices. The polar regions receive their maximum insolation amounts during their summer solstices when the day is continuous.

The above pattern of insolation distribution is slightly modified for the earth's surface primarily because of the effect of the atmosphere. The atmosphere absorbs, reflects, scatters and re-radiates solar energy. About 18% of insolation is absorbed directly by ozone and water vapour. Ozone absorbs all ultraviolet radiation below 0.29 μm. Absorption of radiation by water vapour is greatest between 0.9 μm and 2.1 μm. CO_2 absorbs radiation with wavelengths greater than 4 μm. Cloud cover impedes the penetration of insolation. The amount of radiation reflected by cloud depends on not only cloud amount and thickness but also on the type of cloud (see Table 3.1). On average about 25% of incoming solar radiation is reflected back to space by clouds. Radiation is also reflected by the earth's surface. Again, values of albedo vary with the type of surface (see Table 3.2). In general, light coloured or dry surfaces reflect more radiation

Table 3.1 Albedo of various types of cloud (after Sellers, 1965, and Barry and Chorley, 1976)

Type of cloud	Albedo (%)
Cumuliform	70–90
Cumulonimbus: large and thick	92
Stratus (150–300 metres thick)	59–84
Stratus 500 metres thick over ocean	64
Stratus thin over ocean	42
Altostratus	39–59
Cirrostratus	44–50
Cirrus over land	36

Table 3.2 Albedo of various types of surfaces

Surface	Albedo (%)
Dry black soil	14
Moist black soil	8
Bare ground	7–20
Sand	15–25
Forests	3–10
Green fields	3–15
Dry ploughed fields	20–25
Grass	15–30
Fresh snow	80
Old snow	50–70
Ice	50–70
Water, solar altitude > 40°	2–4
Water, solar altitude 5–30°	6–40
Cities	14–18

than dark coloured or wet surfaces. The albedo of most surfaces varies with the wavelength and angle of incidence of the light rays. Most types of soil and vegetation for instance have very low albedo in the ultraviolet increasing in the visible and infrared. The highest albedo of ice is, however, near 0.55 μm with lower values at both shorter and longer wavelengths (Sellers, 1965). Vertical light rays generally give lower albedo than slanting or inclined rays. Hence, the albedo of a given surface is high near sunrise and sunset and low around noon.

Insolation is scattered mainly by air molecules, water vapour, and particulate matter within the atmosphere. The scattering may be upward towards space or downwards towards the earth's surface. About 6% of the insolation reaching the top of the atmosphere is scattered downwards and reaches the surface as diffuse radiation. The shorter wavelengths are affected by *Rayleigh scattering* which occurs when the diameters of the particles are smaller than the wavelengths of the solar radiation. Rayleigh scattering applies to particles of radius less than $10^{-1}\lambda$, notably air molecules. The effect of Rayleigh scattering is best seen when the atmosphere is free of suspended particles. Then the sky light is a brilliant blue. On the other hand, when there are dust and haze particles in the atmosphere the sky light is white. This is because of the *Mie scattering* effect which operates when diameters of particles are larger than the wavelengths of the incident radiation. Scattering is then non-selective and is effective for all wavelengths.

Two other factors which influence the distribution of insolation over the earth's surface are:

1 the distribution of land and water surfaces; and
2 elevation and aspect.

Land and water have different thermal properties and react differently to insolation. Water heats up at a lower rate than land but it loses its heat less readily. Thus, whereas water has a tendency to store the heat it receives, the land, on the other hand, quickly returns it to the atmosphere. These differences in the thermal properties of land and water surfaces help produce what is called the *continentality effect* discussed in the next chapter. They are also responsible for the land and sea breezes experienced in coastal areas and to a large extent account for the nature of Asiatic and other monsoon wind systems (see Chapter 6).

The land and water surfaces behave differently to insolation for several reasons. First, the albedo of the land surface is generally larger than that of water surface. The albedo for land surfaces generally varies from 8 to 40% whereas for a calm water surface the albedo is generally less than that although it can be more than 50% when solar elevation angle is about 15°. Second, water surface is transparent so that the rays of the sun can penetrate deeper than is the case with the relatively opaque land surface. Third, heat transfer in water is mainly by convection, a more efficient and quicker method of heat transfer than the slow conduction process by which heat is transferred within land. Fourth, the specific heat of water is greater than that of land. Water must absorb five times as much heat energy to raise its temperature by the same amount as a comparable mass of dry soil. Also for equal volumes of water and soil, the heat capacity of water exceeds that of the soil two fold. Finally, because water is always easily available for evaporation on the water surface, evaporation there is continuous whereas on land evaporation only takes place if the soil is moist. Since evaporation is a cooling process involving utilization of energy it must be considered in any comparison of the thermal properties of land and water surfaces.

Elevation and aspect do exercise some control over insolation distribution on the earth's surface particularly at the micro- or local scale. Insolation values at high elevation under clear skies are generally higher than those in locations near the sea level in the same environment. This is because the smaller mass of air above locations at high altitude ensures less interference with insolation by the atmosphere. Aspect refers to the direction faced by a given slope. Some slopes are more exposed to the sun than others. In the middle and high latitudes, poleward facing slopes generally receive less radiation than slopes facing the equator. In the alpine valleys of Europe, for example, settlements and cultivation

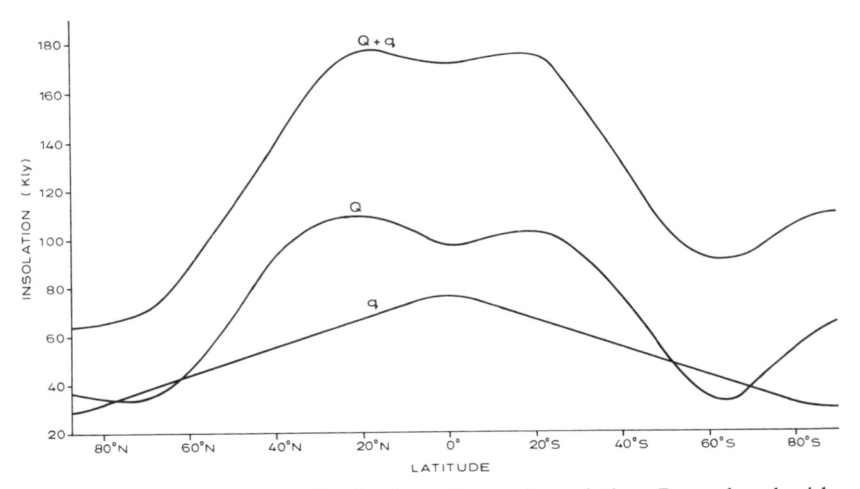

Fig. 3.3. Annual latitudinal distribution of annual insolation. Reproduced with permission from Sellers, *Physical Climatology*, 1965. Published by The University of Chicago Press

are noticeably concentrated on the southward facing slopes (the 'adret' or sunny side) whilst northward facing slopes (the 'ubac' or shaded side) are left under forest.

The average annual latitudinal distribution of insolation over the earth is depicted in Fig. 3.3. The graph indicates that the highest amounts of insolation are received in the subtropical zones which have slightly higher values than the more cloudy equatorial zone. Insolation values decrease poleward, reaching their minima around latitudes 70–80° north and latitudes 60–70° south of the equator. The distribution of annual insolation over the globe according to Budyko is shown in Fig. 3.4. The highest values of more than 200 kly/year

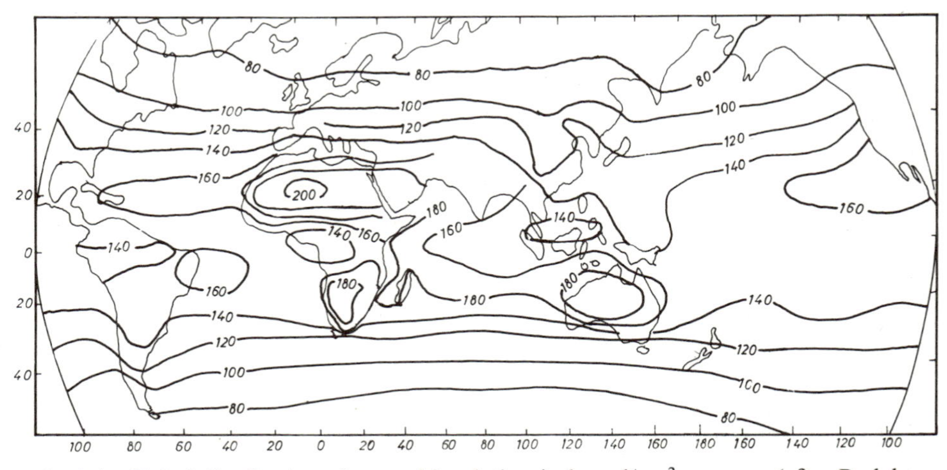

Fig. 3.4. Global distribution of annual insolation in kg cal/cm² per year (after Budyko, 1958)

are found over the major deserts of the world where as much as 80% of the solar radiation incident on top of the atmosphere during the year reaches the ground. Values of less than 100 kly/year occur poleward of latitude 40° over the oceans and of latitudes 50° over the continents as well as around the equator in central Africa.

The distribution of December and June insolation values as computed by Budyko are shown in Fig. 3.5. In December, insolation values are higher in the southern hemisphere than in the north while the reverse situation occurs in June. In December the highest insolation values of more than 18 kg cal/cm²

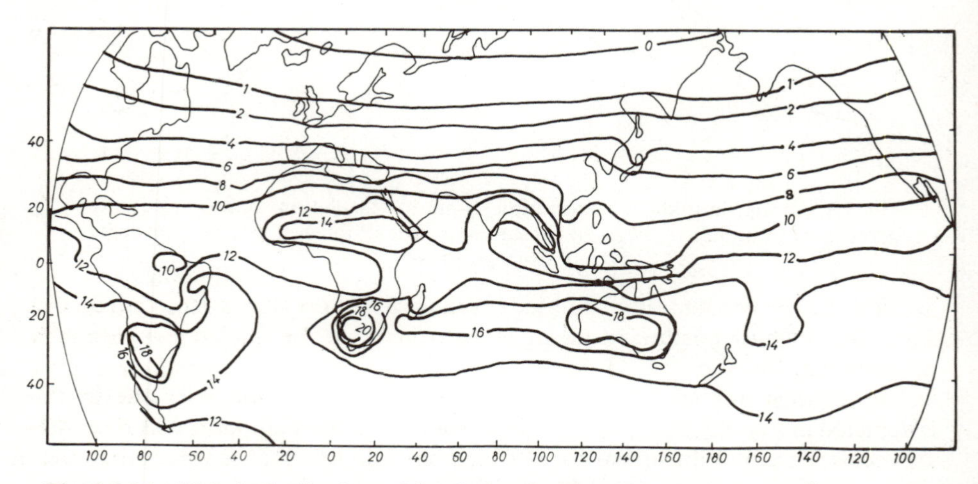

Fig. 3.5 (a). Global distribution of insolation in December in kg cal/cm² per month (after Budyko, 1958)

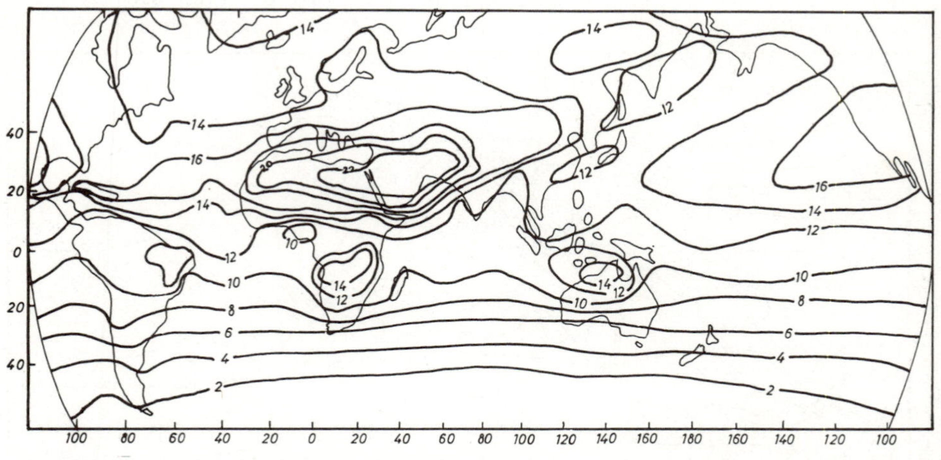

Fig. 3.5 (b). Global distribution of insolation in June in kg cal/cm² per month (after Budyko, 1958)

occur in southern Africa, central Australia and South America. Except for the moderately high values occurring over the savanna belt of West Africa and the Sudan, insolation values generally decrease steadily towards the north pole. Beyond the Arctic circle insolation is zero as this area is continuously under darkness. In June, the highest insolation amounts occur in the subtropical zone of the northern hemisphere. Insolation values decrease slowly towards the north pole but more rapidly towards the south pole. Beyond the Arctic circle insolation values are around 14 kg cal/cm^2. On the other hand, in the southern hemisphere insolation values are less than 2 kg cal/cm^2 poleward of latitude 40 °S. During this period, the area beyond the Antarctic circle is continuously in the dark.

Terrestrial radiation

The earth's surface when heated by the absorption of solar radiation becomes a source of long-wave radiation. The average temperature of the earth's surface is only 10 °C compared to 6000 °C for the sun. As indicated by Wien's displacement law given earlier, most of the radiation emitted by the earth is in the infrared spectral range from 4 μm to 100 μm with a peak near 10 μm. Terrestrial radiation is also called nocturnal radiation since it is the major radiative source of energy at night. It is important to note, however, that infrared radiation is not necessarily terrestrial since atmospheric constituents also radiate energy in the infrared wavelengths. Second, infrared radiation occurs during both day and night. It is only dominant at night because solar radiation is then cut off.

It is usually assumed that the earth's surface emits and absorbs energy as a grey body in the infrared wavelength so that the flux of terrestrial radiation ($I_\uparrow$) is given by the equation of the form

$$I_\uparrow = \varepsilon \sigma T^4 \tag{3.3}$$

where ε is the infrared emissivity, σ is the Stefan–Boltzmann constant and T is the absolute temperature of the earth. Typical infrared emissivities for various surfaces are given in Table 3.3. The emissivity of a black body is 1.0. Infrared

Table 3.3 Infrared emissivities for various surfaces

Surface	Emissivity (%)
Water	92–96
Fresh snow	82–99.5
Dry sand	89–90
Wet sand	95
Moist bare ground	95–98
Desert	90–91
Dry high grass	90
Shrubs	90
Woodland	90
Human skin	95

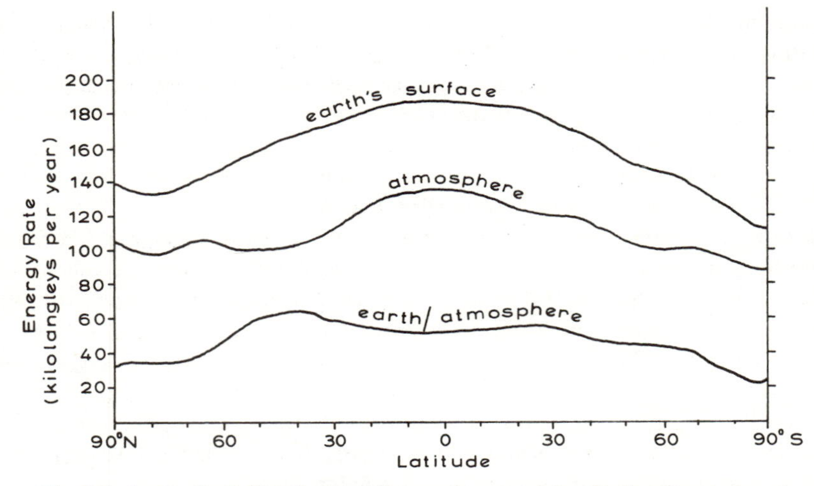

Fig. 3.6. Latitudinal distribution of annual terrestrial radiation. Reproduced with permission from Sellers, *Physical Climatology*, 1965. Published by The University of Chicago Press

emissivity is equivalent to infrared absorptivity, i.e. $(1 - \alpha_1)$ where α_1 is infrared albedo.

The average latitudinal distribution of infrared terrestrial radiation $(I_{\uparrow})$ is shown in Fig. 3.6. Values of infrared terrestrial radiation are highest in the low latitudes where values are about 280 kg cal/cm^2 per year decreasing to less than 150 kg cal/cm^2 per year around the north pole and less than 120 kg cal/cm^2 per year around the south pole.

Atmospheric radiation

Like the earth, the atmosphere absorbs and emits radiant energy. Although the atmosphere is nearly transparent to short-wave radiation it is highly absorbent to infrared radiation. The principal absorbers of infrared radiation amongst atmospheric constituents are water vapour (5.3 μm to 7.7 μm and beyond 20 μm), ozone (9.4 μm to 9.8 μm), carbon dioxide (13.1 μm to 16.9 μm), and clouds which absorb radiation at all wavelengths. Whereas the atmosphere absorbs only 24% of incoming solar radiation which is short wave, only 9% of infrared terrestrial radiation escapes directly to space mainly through the so-called atmospheric window made of wavelengths in the range 8.5–11.0 μm. The remaining 91% is absorbed by the atmosphere. This opaqueness of the atmosphere to infrared radiation relative to its transparency to short-wave radiation is usually referred to as the *greenhouse effect*. In other words, the atmosphere acts like the glass in a greenhouse admitting solar radiation but not allowing terrestrial radiation to escape to space. The atmosphere re-radiates the absorbed terrestrial and solar radiation partly to space and partly back to the earth's surface (counter radiation). Clouds are particularly effective radiators acting like black bodies. Cloudiness and cloud top temperatures can therefore be mapped from satellites both day and night using infrared sensors. The radiative cooling of cloud layers is estimated to average about 1.5 °C per day (Barry and Chorley, 1976). It is believed that without atmospheric counter radiation, the earth's surface would be 30–40 °C cooler than it is now.

Radiation balance

By radiation balance we mean the difference between the amounts of radiation which are absorbed and emitted by a given body or surface. In general, the radiation balance at the earth's surface is positive by day and negative at night. Also over the year as a whole, the radiation balance at the earth's surface is positive while that of the atmosphere is negative. For the whole earth–atmosphere system the balance is positive between latitudes 30 °S and 40 °N and negative elsewhere. These patterns of radiation balance have implications for the general circulation of the atmosphere as discussed latter in Chapter 5.

The radiation balance equation is of the form

$$R = (Q + q)(1 - \alpha) + I_{\downarrow} - I_{\uparrow} \tag{3.4}$$

where R is the radiation balance or the net radiation, $(Q + q)$ is the sum of direct and diffuse solar radiation incident on the earth's surface, α is the surface albedo, $I_{\downarrow}$ counter radiation from the atmosphere and $I_{\uparrow}$ is terrestrial radiation. The solar energy incident on top of the earth's atmosphere is about 263 kly per year. Only 169 kly of energy are absorbed by the earth–atmosphere system, the remaining 94 kly being reflected back to space. The latter amount constitutes about 36% of the total energy incident on top of the earth's atmosphere and is referred to as the *planetary albedo*. As shown in Tables 3.4 and 3.5, the effective outgoing radiation from the earth–atmosphere amounts to 169 kly a year. This means that for the earth–atmosphere system radiation balance or net radiation is zero (see Table 3.6). The atmosphere absorbs 45 kly of energy per year while the effective outgoing radiation from the atmosphere is 117 kly. This leaves a negative radiation balance of 72 kly. For the earth's surface the balance is positive and amounts to 72 kly. Although the radiation balance

Table 3.4 Global disposition of solar radiation incident on top of the atmosphere during an average year in kly per year (after Sellers, 1965)

Solar energy incident on top of atmosphere	263
Reflected by clouds	63
Reflected by molecules, dust, and water vapour	15
Total reflected by the atmosphere	78
Reflection from the earth's surface	16
Total reflected by the earth–atmosphere system	94
Absorbed by clouds	7
Absorbed by molecules, dust, and water vapour	38
Total absorbed by the atmosphere	45
Absorbed by the earth's surface	124
Total absorbed by the earth–atmosphere system	169

Table 3.5 Global disposition of infrared radiation in the earth–atmosphere system during an average year in kly per year (after Sellers, 1965)

Infrared radiation emitted by the earth's surface	258
Lost to space	220
Absorbed by the atmosphere	238
Infrared radiation emitted by the atmosphere	355
Lost to space	149
Absorbed by the earth's surface as counter radiation	206
Effective outgoing radiation from the earth's surface	52
Effective outgoing radiation from the atmosphere	117
Effective outgoing radiation from the earth–atmosphere system	169

Table 3.6 The radiation balance during an average year in kly per year

	Gain	Loss	Net
The earth's surface	124	52	72
The atmosphere	45	117	−72
Earth–atmosphere	169	169	0

averaged over the year are −72 kly, 72 kly, and zero for the atmosphere, the earth and the earth–atmosphere system respectively, there are seasonal and annual variations in any given latitudinal zone. As shown in Fig. 3.7, the atmosphere is uniformly a radiative sink at all latitudes while the earth's surface is uniformly a heat source except near the poles. To prevent the earth's surface from warming and the atmosphere from cooling, surplus energy is transferred

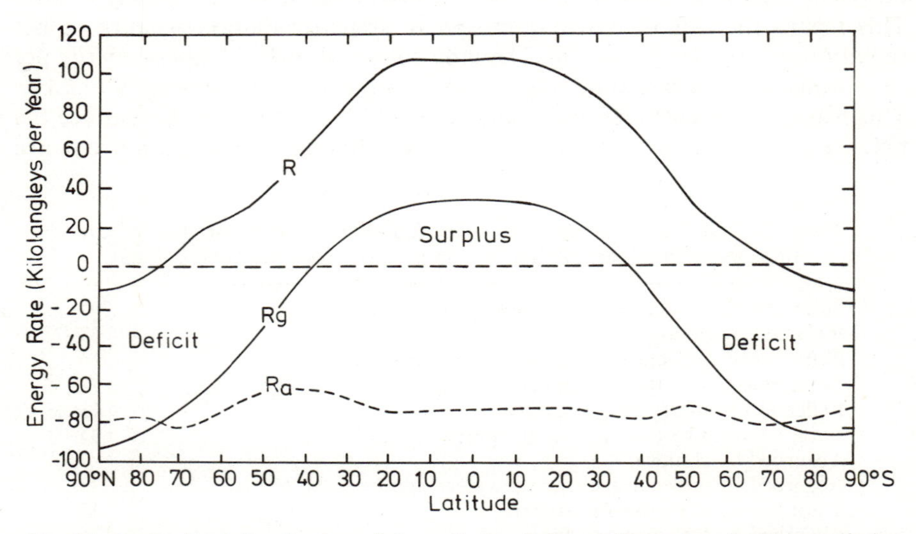

Fig. 3.7. Latitudinal distribution of the radiation balance. Reproduced with permission from Sellers, *Physical Climatology*, 1965. Published by The University of Chicago Press

from the earth's surface to the atmosphere to make up for the deficit there. This vertical exchange of energy takes place mainly by:

1 evaporation of water from the earth's surface and the condensation of the vapour in the atmosphere to release the latent heat;
2 the conduction of sensible heat from the earth's surface to the atmosphere; and
3 convection, i.e. turbulent diffusion of heat into the atmosphere from the earth's surface.

Poleward of about 40° latitude, the radiative deficit of the atmosphere exceeds the surplus of the surface so that the radiation balance of the earth–atmosphere system in these areas in negative. Conversely, in the low latitudes towards the equator of 40° latitudes the global radiation balance is positive. To prevent the tropics from getting warmer and the poles from getting colder, there is a meridional transfer of energy from the low to the middle and high latitudes. This horizontal heat exchange over the earth's surface is induced also partly by the differential heating of the continents and the oceans and is carried out mainly by:

1 the poleward transfer of sensible heat by the atmospheric circulation and ocean currents from the low latitudes;
2 the release of latent heat when water vapour carried poleward from the low latitudes condenses in the atmosphere.

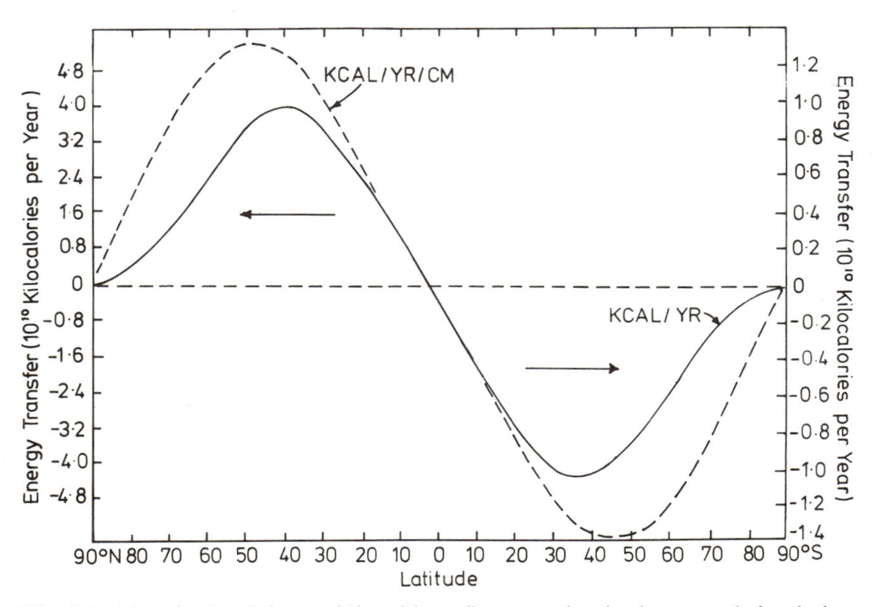

Fig. 3.8. Magnitude of the meridional heat flux to maintain the general circulation of the atmosphere balance. Reproduced with permission from Sellers, *Physical Climatology*, 1965. Published by The University of Chicago Press

Thus, we have a two-way heat transfer within the earth–atmosphere system from the earth's surface to the atmosphere and from the equator to the poles. This transfer is the *raison d'être* for the general circulation of the atmosphere and must occur in such a way that no part of the earth–atmosphere system warms or cools noticeably over a period of a year. The magnitude of the meridional heat flux required to maintain this balance is shown in Fig. 3.8. As shown in this diagram, the meridional heat flux is highest between latitudes 40° and 50° in each hemisphere and is slightly higher in the southern hemisphere than in the north.

The energy budget of the earth

Energy balance or budget is a concept used in climatology to relate the net radiation flux to latent heat and sensible heat transfer among others in the equation of the form

$$R = LE + H + G + \Delta f + P \tag{3.5}$$

where R is the radiation balance or net radiation, LE is latent heat of evaporation, H is sensible heat, Δf is net horizontal advection of heat by currents, G

Table 3.7 Mean latitudinal values of components of the energy balance equation for the earth's surface in kly per year (after Sellers, 1965)

Latitudinal zone	Oceans				Land			Earth			
	R	LE	H	Δf	R	LE	H	R	LE	H	ΔF
80–90 °N								−9	3	−10	−2
70–80 °N								1	9	−1	−7
60–70 °N	23	23	16	−26	20	14	6	21	20	10	−9
50–60 °N	29	39	16	−26	30	19	11	30	28	14	−12
40–50 °N	51	53	14	−16	45	24	21	48	38	17	−7
30–40 °N	83	86	13	−16	60	23	37	73	59	24	−10
20–30 °N	113	105	9	−1	69	20	49	96	73	24	−1
10–20 °N	119	99	6	14	71	29	42	106	81	16	9
0–10 °N	115	80	4	31	72	48	24	105	72	11	22
0–90 °N								72	55	16	1
0–10 °S	115	84	4	27	72	50	22	105	76	10	19
10–20 °S	113	104	5	4	73	41	32	104	90	11	3
20–30 °S	101	100	7	−6	70	28	42	94	83	16	−5
30–40 °S	82	80	8	−6	62	28	34	80	74	11	−5
40–50 °S	57	55	9	−7	41	21	20	56	53	10	−7
50–60 °S	28	31	10	−13	31	20	11	28	31	11	−14
60–70 °S								13	10	11	−8
70–80 °S								−2	3	−4	−1
80–90 °S								−11	0	−11	0
0–90 °S								72	62	11	−1
Globe	82	74	8	0	49	25	24	72	59	13	0

is heat transferred into or out of storage and P is energy used for photosynthesis. The amount of energy used for photosynthesis is very small (about 5% of the net radiation). Over land surfaces, Δf is negligibly small while in the annual heat balance equation the heat storage term may be neglected or regarded as constant. This is because the heat stored in spring and summer is released in the autumn and winter. Similarly heat stored in the morning and early afternoon is lost in later afternoon and at night. Thus for oceans the energy balance equation may be written

$$R = LE + H + \Delta f \tag{3.6}$$

and for the land surfaces it may be written

$$R = LE + H. \tag{3.7}$$

Table 3.7 shows the mean latitudinal values of the components of the energy balance equation for the earth's surface while Table 3.8 shows the annual energy balance of the ocean and continents. On both the ocean and the land, the highest amounts of net radiation are found in the tropics. In the low latitudes, net radiation values over the oceans are higher than those over land surfaces. This is because of the relatively high albedo of the land surface in this area and the fact that the land areas in this zone are predominantly deserts with little or no cloud cover. The absorbed solar radiation in this zone is therefore less and the effective outgoing radiation greater over land than over oceans. Poleward of 50° latitude in both hemispheres net radiation values over land and ocean surfaces are about the same. This is because in these regions the albedo of the water surface is

Table 3.8 Annual energy balance of the oceans and continents in kly per year (after Sellers, 1965)

Area	R	LE	H	Δf	H/LE
Europe	39	24	15	0	0.62
Asia	47	22	25	0	1.14
North America	40	23	17	0	0.74
South America	70	45	25	0	0.56
Africa	68	26	42	0	1.61
Australia	70	22	48	0	2.18
Antarctica	−11	0	−11	0	
All land	49	25	24	0	0.96
Atlantic Ocean	82	72	8	2	0.11
Indian Ocean	85	77	7	1	0.09
Pacific Ocean	86	78	8	0	0.10
Arctic Ocean	−4	5	−5	−4	−1.00
All oceans	82	74	8	0	0.11
Northern hemisphere	72	55	16	1	0.29
Southern hemisphere	72	62	11	−1	0.18
Globe	72	59	13	0	0.22

relatively high owing to the low solar altitude. Near the poles net radiation is negative since the effective outgoing radiation exceeds the small amount of radiation absorbed by the highly reflective ice and snow covered surfaces. For the earth as a whole the net radiation is about 70% greater over the oceans than over land (Sellers, 1965).

On the land, the latent heat flux (*LE*) is highest at the equator and generally decreases poleward. But the latent heat flux over the oceans is highest in the subtropics between 10° and 30° latitudes and decreases both equatorward and poleward. The latent heat flux over the oceans is generally more than twice that over land where water is less easily available for evaporation. For the earth as a whole evaporation rates from land are only about one-third of those from the oceans.

The sensible heat flux or turbulent heat exchange increases from the equator towards the poles over the oceans. In contrast, sensible heat flux from the land surfaces is highest in the subtropics and decreases both poleward and equatorward. Poleward of 70° latitude, there is a downward or negative sensible heat flux because the earth's surface is usually colder than the overlying air. For the earth as a whole the transfer of sensible heat from the land areas exceeds that from the oceans by a factor of three (Sellers, 1965).

Table 3.8 shows that 90% of the net radiation of the oceans is used to evaporate water and the remaining 10% is used to warm the air by conduction and convection. In contrast latent heat flux and sensible heat flux are equally important forms of heat loss on the continents. For the earth as a whole, latent heat flux accounts for 82% of the net radiation and turbulent heat exchange accounts for 18%.

The mean latitudinal values of the heat budget components shown in Table 3.7 and the mean values of the heat budget components by continents and oceans

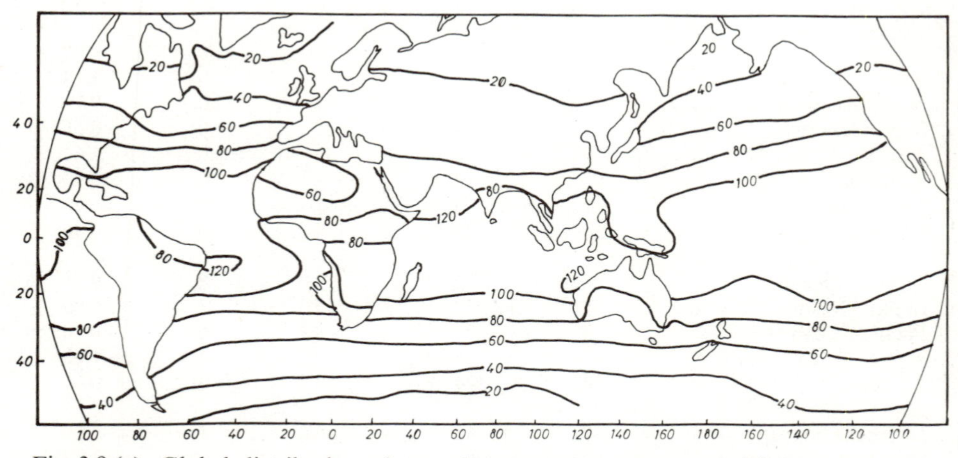

Fig. 3.9 (a). Global distribution of annual net radiation in kg cal/cm² per year (after Budyko, 1958)

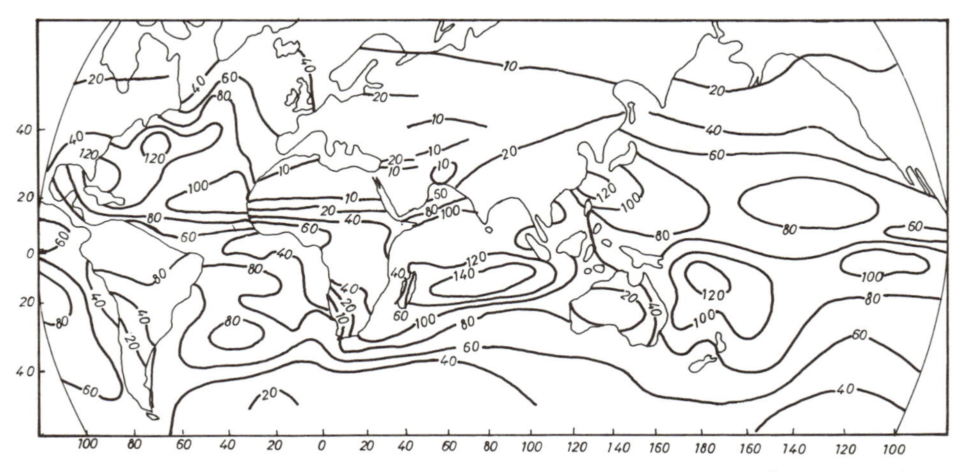

Fig. 3.9 (b). Global distribution of annual latent heat flux in kg cal/cm^2 per year (after Budyko, 1958)

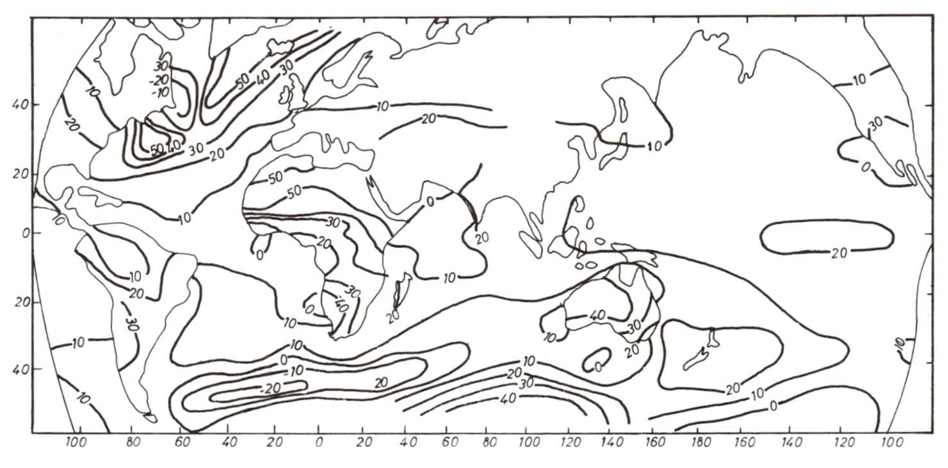

Fig. 3.9 (c). Global distribution of annual sensible heat flux in kg cal/cm^2 per year (after Budyko, 1958)

shown in Table 3.8 and discussed above conceal great spatial variations which can only be portrayed on world maps. The global distributions of annual net radiation, latent heat flux and sensible heat flux are presented in Fig. 3.9. Fig. 3.9(a) shows that:

1 the annual net radiation values are highest in the low latitudes and decrease poleward from about 25° latitude;
2 net radiation values are somewhat higher on the oceans than on the continents in the same latitudes mainly because of greater absorbtion of radiation over oceans and less effective outgoing radiation;

3 net radiation values are rather lower in arid continental areas than in humid lands because of greater effective outgoing radiation in arid land areas under relatively clear skies.

Figs. 3.9(b) and 3.9(c) show that both latent heat and sensible heat fluxes are distributed differently over land and water surfaces. Latent heat flux is highest over water surfaces in the low latitudes where it exceeds 120 kly a year. It is least in arid areas where there is little available energy. The largest sensible heat exchange takes place in the tropical deserts where more than 60 kly a year is transferred to the atmosphere. The lowest amount occurs in areas of cold currents where there is negative sensible heat as warm continental air masses move off shore over cold currents transferring energy to the oceans (Barry and Chorley, 1976).

Because of the dearth of radiation data over most parts of the world, we have to rely on estimates arrived at using various formulae. Although satellites are now providing a top view of the radiation exchanges in the earth–atmosphere system some uncertainties are still to be resolved. In the section below, the measurement of radiation is briefly discussed. Measurement of the various components of the radiation balance equation are required for more detailed and accurate mapping of the distribution of these parameters over the earth's surface or part thereof. Most detailed mapping of radiation balance components at regional or national level is still based on estimated values derived from various empirical equations. Such maps though detailed cannot be regarded as more accurate than maps of global distribution of these parameters which are usually based on actual measurements.

Measurement of radiation

Many instruments are available for measuring the components of the radiation balance but they are generally expensive compared to the more common meteorological instruments like the thermometer or the rain gauge. There are five basic types of radiation measuring instruments. These are:

1 pyrheliometers which measure the solar intensity, i.e. the direct beam solar radiation at normal incidence. Pyrheliometers are expensive but are the most accurate of all radiation measuring instruments. They are therefore commonly used as calibration standards.
2 pyranometers which measure the total short-wave radiation from sky incident on a horizontal surface at the ground.
3 pyrgeometers which measure infrared radiation.
4 pyrradiometers which measure both infrared and solar radiation together.
5 net radiometers which measure the net radiation or the radiation balance.

Some of these instruments measure different parameters when shaded as indicated in Table 3.9. Thus, it is possible to measure the solar intensity using two pyranometers, one shaded and the other unshaded, or by alternately shading and unshading a single pyranometer. A pyrradiometer or net radiometer

Table 3.9 Measurements of some radiation components using shaded and unshaded instruments

	Pyranometer	Pyrradiometer	Net radiometer
Unshaded	$Q+q$	$Q+q+I_{\downarrow}$	R
Shaded	q	$q+I_{\downarrow}$	$R-Q$
Unshaded–Shaded	Q	Q	Q

Q, direct beam solar radiation; q, diffuse radiation; $I_{\downarrow}$, counter radiation from the atmosphere; R, net radiation.

can be similarly used to obtain the value of the direct beam solar radiation on a horizontal surface (Q). The solar intensity (Qn) can then be obtained by dividing Q by the cosine of the solar zenith angle. Albedo, the reflectivity of a given surface, can also be measured using two pyranometers one of which faces downward or with a single pyranometer which is inverted periodically. The upward facing pyranometer will record the total solar radiation ($Q+q$) while the downward facing pyranometer records only the reflected radiation $(Q+q)\alpha$. The albedo is then obtained by taking the ratio of the two reading. Thus, pyranometers are very useful and versatile. Besides measuring the total solar

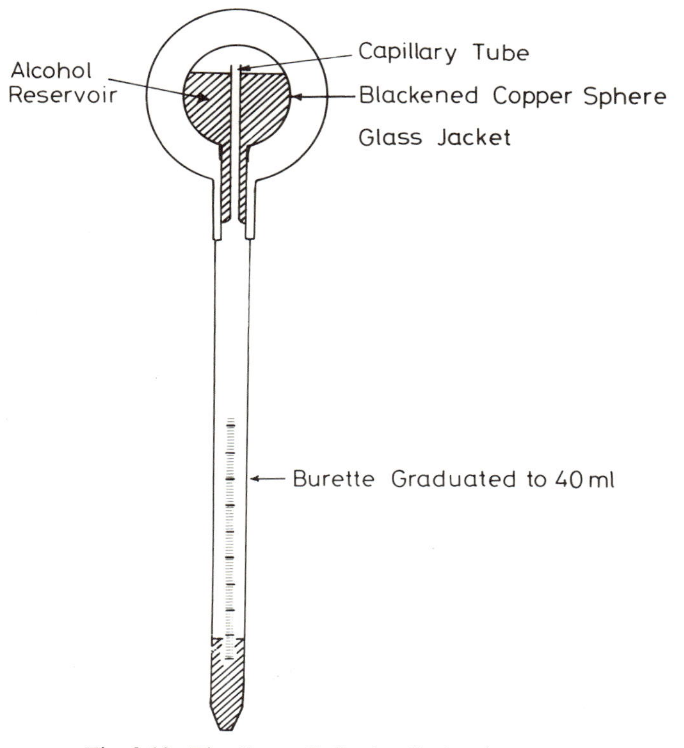

Fig. 3.10. The Gunn–Bellani radiation integrator

radiation falling on a horizontal surface they can be used to measure the total solar radiation falling on an inclined surface, the solar intensity, albedo, and the diffuse sky radiation (q).

Because the above radiation measuring instruments are sophisticated and expensive they are not commonly used in most parts of the tropics. Instead insolation in these areas is usually estimated using the Gunn Bellani radiation integrator or the Campbell–Stokes sunshine recorder. The Gunn Bellani radiation integrator can be described as a spherical type of pyranometer. It has two concentric glass spheres mounted on the end of a burette graduated in millilitres up to 40 ml. The inner sphere is covered by a black-coated copper shell while a reservoir of pure ethyl alcohol within the inner sphere is open to the burette through a small capillary tube (see Fig. 3.10). As the black inner sphere absorbs radiation and is heated some of the liquid evaporates only to condense again at the bottom of the burette. The amount of distillate in a given time is

Fig. 3.11. The Campbell–Stokes sunshine recorder

directly related to the amount of solar-radiation absorbed. The instrument which is put in a hole in the ground with the spherical glass exposed to insolation may be read hourly or daily. After each reading, the instrument is reset by inverting it so that the alcohol can return to the reservoir.

The Campbell–Stokes sunshine recorder is used to measure the duration of bright sunshine. It consists of a glass sphere which focuses the sun's rays to a sensitized card graduated in hours and held in a metal half-bowl with which the sphere is concentric (see Fig. 3.11). The instrument is usually mounted on a concrete pillar of about 1.5 m height above the ground. Bright sunshine burns a path along the sensitized card while cloudy periods are blank. The total duration of sunshine for the day is obtained by measuring the total length of the brown trace on the card. Values of insolation are closely related to the duration of bright sunshine and can be estimated using a regression equation of the form

$$Q = Q_0(a + bn/N) \tag{3.8}$$

where Q is measured insolation, Q_0 is possible insolation, i.e. radiation incident on top of the atmosphere obtainable from Smithsonian Meteorological Tables, n is measured duration of sunshine and N is possible duration of sunshine obtainable from Smithsonian Meteorological Tables.

The above equation has been used by various authors to estimate and map values of solar radiation for various countries and regions of the world (see for example Davies, 1966; Ojo, 1970). Similar empirical equations abound for estimating values of net radiation (see Penman, 1948). The reliability of these estimates is, however, questionable particularly when equations derived in one area are used in another. Such estimates should therefore be treated with caution. The existing network of radiation measuring stations in the world particularly in the tropics needs to be considerably improved in view of the important role of radiation in our climatic and biotic environment and as the ultimate source of energy for the use of man.

References

Barry, R. G. and Chorley, R. J. (1976). *Atmosphere, Weather and Climate* (3rd edn). Methuen, London.

Budyko, M. I. (1958). *The heat balance of the earth's surface.* (Translated from Russian by N. A. Stepanova) U. S. Department of Commerce, Washington, D. C.

Davies, J. A. (1966). Solar radiation estimates for Nigeria. *Nigeria Geog. J.* **8**, No. 1, 85–100.

Ojo, S. O. (1970). The seasonal march of the spatial patterns of global and net radiation in West Africa. *J. of Tropical Geog.* **30**, 48–62.

Penman, H. L. (1948). Natural evaporation from open water, bare soil and grass. *Proceedings Royal Society Series A* **193**, 120–145.

Sellers, W. D. (1965). *Physical Climatology.* University of Chicago Press, Chicago.

CHAPTER 4

Temperatures

Temperature and its measurement

Apart from precipitation, temperature is probably the most talked about weather element. Temperature can be defined in terms of movement of molecules such that the more rapid the movement the higher the temperature. More usually, it is defined in relative terms on the basis of the degree of heat a body has. Temperature is the condition that determines the flow of heat from one substance to another. Heat moves from a body having a higher temperature to a body with a lower temperature. The temperature of a body is determined by the balance between incoming and outgoing radiation and its transformation into sensible and latent heat among others as shown in the radiation balance and the energy budget equations discussed earlier in Chapter 3.

The temperature of a body is therefore its degree of hotness as measured by a thermometer. Various scales are used to express temperatures. These include the Fahrenheit, Centigrade, and the Kelvin or absolute temperature scales. In most countries, temperatures are now expressed on the Centigrade or Celsius scale and in some applications the Kelvin or absolute temperature scale is used. Temperatures on Fahrenheit (F) scale can be converted to Celsius (C) using the simple equation of the form

$$C = \tfrac{5}{9}(F - 32). \qquad (4.1)$$

The zero on the Kelvin or absolute temperature scale is the temperature at which a gas would theoretically cease to exert any pressure and that temperature is $-273\ ^{\circ}\mathrm{C}$. Centigrade temperatures may therefore be converted to °A (or K) by adding 273 to them. Alternatively, Kelvin or absolute temperatures can be converted to °C by substracting 273 from them.

There are different types of thermometers depending on the measuring element, i.e. the medium used to measure changes in temperatures. There are electrical thermometers, resistance thermometers, gas thermometers, thermocouples, mercury-in-glass thermometers and alcohol thermometers. All these are described in elementary physics textbooks such as that by Nelon and Parker (1970). Meteorologists measure the highest and lowest temperatures reached

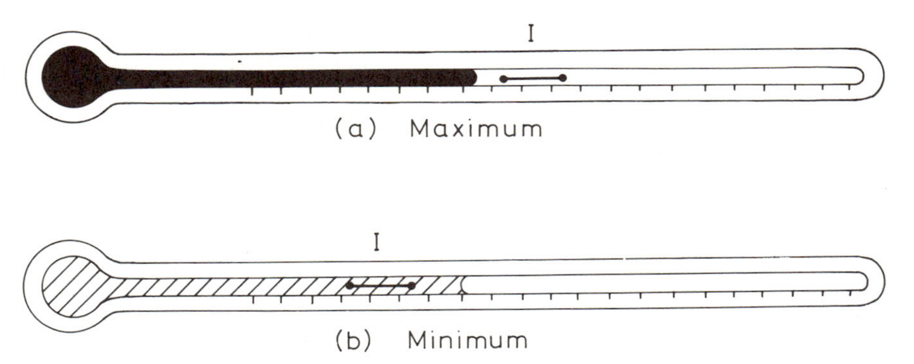

Fig. 4.1. Maximum and minimum thermometers

by the air in a day using the maximum and the minimum thermometers (see Fig. 4.1). These are kept at a height of about 1.5 metres above the ground in a white wooden-louvred shelter called the Stevenson screen.

The maximum thermometer is a mercury-in-glass thermometer containing a small glass index which the mercury pushes along when air temperature rises but leaves behind when the temperature falls. The maximum temperature is shown by the end of the index nearer the mercury. After observation, the index is brought back to the mercury by tilting the thermometer. The minimum thermometer is an alcohol-in-glass thermometer in which when the temperature rises the alcohol expands and flows past the index and when the temperature falls the alcohol contracts and drags the index back because of its surface tension.

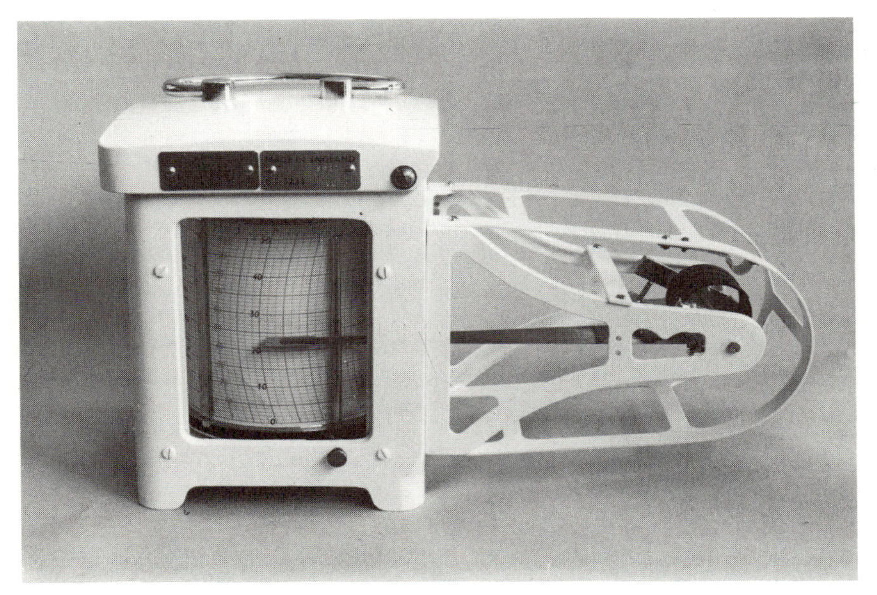

Fig. 4.2. A bimetallic thermograph

The end of the index nearer the meniscus shows the minimum temperature and the instrument is reset by tilting. The air temperature at any given time can be read off an ordinary mercury-in-glass thermometer with or without an index. Air temperature can also be continuously measured with the aid of a self-recording thermometer known as a thermograph of which there are five types (WMO, 1971). Perhaps the most commonly used thermograph is the bimetallic one in which a bimetal strip is wound into a spiral such that the spiral uncoils as the air temperature rises. The movement of the spiral is used to operate a pen which traces a graph of the air temperature (see Fig. 4.2). The thermograph is also kept within a Stevenson screen.

Spatial variations in temperature

Air temperature varies from place to place and over time at a given location. The distribution of temperature over a given area is normally shown by isotherms while the variation of temperature in time is shown by some form of graph. Several factors influence the distribution of temperature over the earth or part thereof. They include the amount of insolation receipt, nature of the surface, distance from water bodies, relief, nature of the prevailing winds, and ocean currents.

Latitude is the primary control of the amount of insolation that a given place would receive. This is because the astronomical variation of insolation is a function of latitude. The angle of incidence of solar rays and the length of day at any given place are all determined by the latitudinal location of such a place. As mentioned in Chapter 3, cloud amount and other atmospheric constituents like aerosols and CO_2 also affect the amount of solar energy that reaches the earth's surface. The nature of the surface under consideration is also of significance since it will determine the values of albedo and specific heat. If albedo is high, less radiation will be absorbed by the surface to raise its temperature. Similarly, if the specific heat of the surface is high, more energy will have to be absorbed by the surface to increase its temperature. The specific heat of sea water is for instance, about 0.94 while that of granite is 0.2. In general water absorbs five times as much heat in order to increase its temperature by the same amount as the earth.

Distance from water bodies influences air temperature because of the basic differences in the thermal characteristics of land and water surfaces (see Chapter 3). These differences help produce the continentality effect in which the land surface heats and cools more rapidly than the water surface. The consequences of this fact are as follows.

1 Over the land, the lag between periods of maximum and minimum surface temperatures is only one month. Over the ocean and at coastal locations the lag is as much as two months.
2 The annual range in temperature is less at coastal locations than for inland locations.
3 Because of greater land area of the northern hemisphere summers there are

Table 4.1 Mean temperatures of the northern and southern hemispheres

	Northern hemisphere	Southern hemisphere
Summer	22.4 °C	17.1 °C
Winter	8.1 °C	9.7 °C

warmer and winters colder than those in the southern hemisphere (see Table 4.1).

Relief has an attenuating effect on temperature primarily because air temperature normally decreases with increasing elevation at the average rate of 0.6 °C per 100 metres. In area of varied topography and slopes, aspect and the degree of exposure of locations are important factors influencing temperature (see Chapter 3).

Elevation is an important factor influencing temperature in the tropics. The relatively thermal uniformity prevailing in the tropics is distorted mainly by the effects of altitude. Large temperature differences over short distances in the tropics are usually due to the effects of variations in altitude. The lapse rate is, however, variable being controlled primarily by elevation and cloudiness. There can also be large differences in temperature conditions between locations on the windward and those on the leeward side of a mountain. In the temperate region, the lapse rate varies considerably with the season being higher in summer than in winter. The lapse rate in the tropics varies less with the season; it is, however, generally larger in the dry season than in the rainy season. This is because under cloudy conditions such as prevail during the rainy season, the effects of radiation are greatly reduced. For the same reason, lapse rates are higher in the dry tropics than near the equator. Whether in the dry or humid tropics, high elevation lowers the temperature and thus brings relief from the oppressive heat of the lowland tropics.

Prevailing winds and ocean currents also influence air temperatures because they can transport or advect 'warmth' or 'cold' from one area to another depending on their thermal characteristics vis-à-vis the area they influence. For instance, coastal areas washed by cold currents have lower temperatures than stations located in similar latitudes but not affected by cold currents. The winter temperatures in northwest Europe are for instance some 11 °C or more above what the temperatures should be judging from the latitudinal location. These increases in temperature in northwest Europe in winter are due to the effect of the warm North Atlantic Drift advecting warmth from the low latitudes towards the middle and upper latitudes.

The pattern of variation of mean surface air temperature over the globe is shown in Figs. 4.3 and 4.4. The main features of these maps are as follows.

1 Air temperatures generally decrease poleward from the equator, a clear evidence of the important role of latitude in influencing insolation and temperatures.

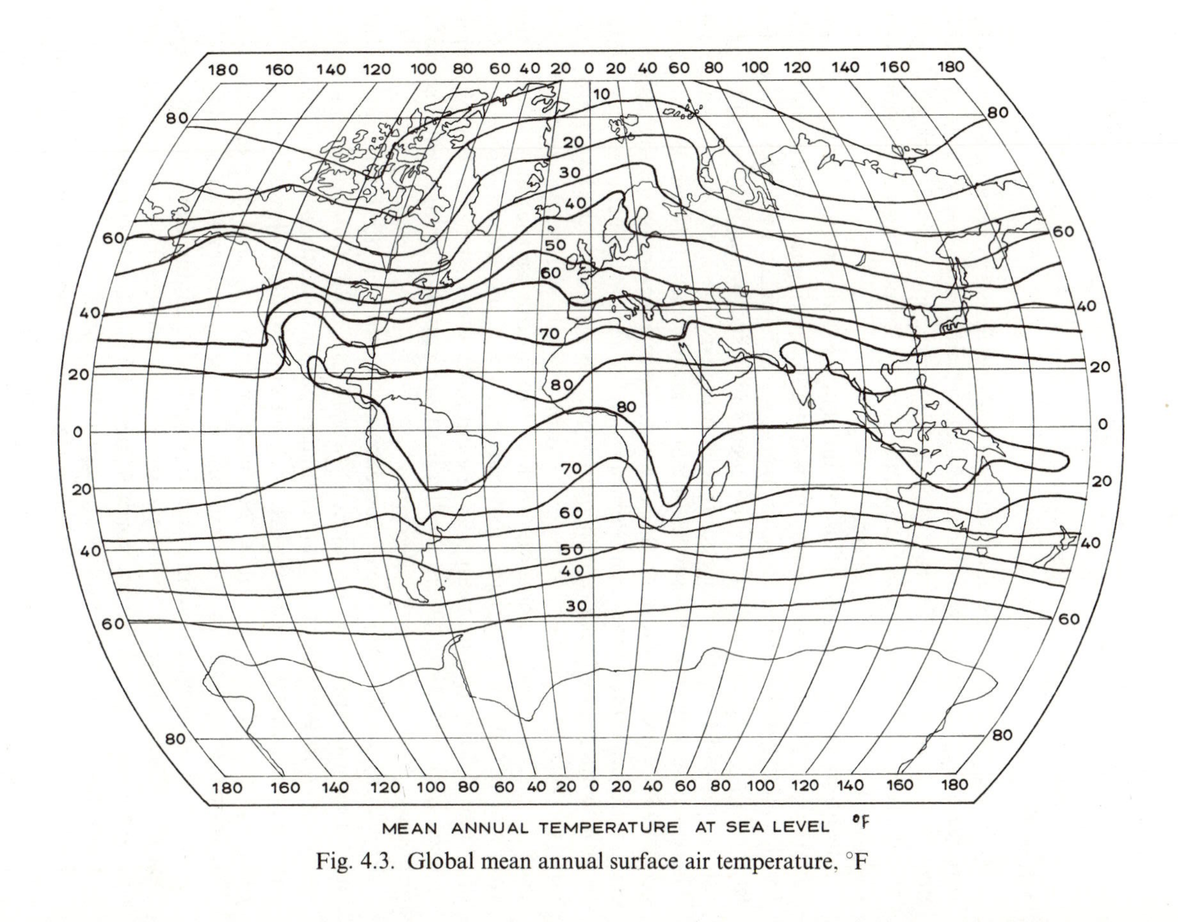

Fig. 4.3. Global mean annual surface air temperature, °F

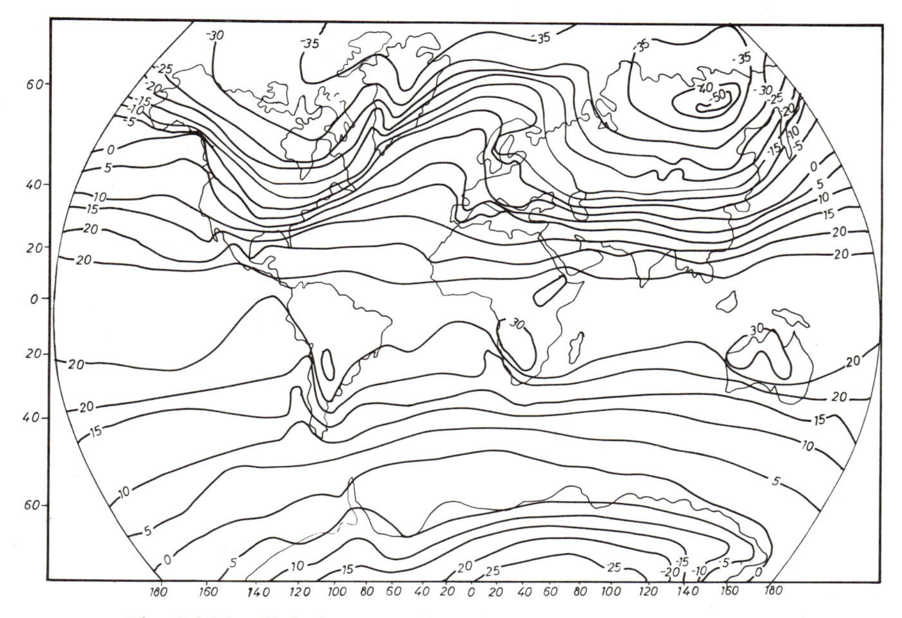

Fig. 4.4 (a). Global mean surface air temperature in January, °C

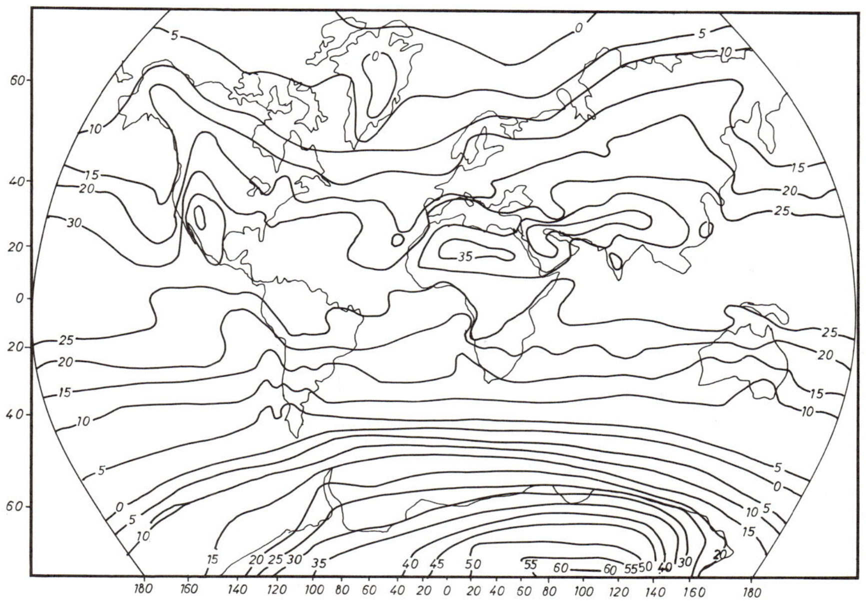

Fig. 4.4 (b). Global mean surface air temperature in July, °C

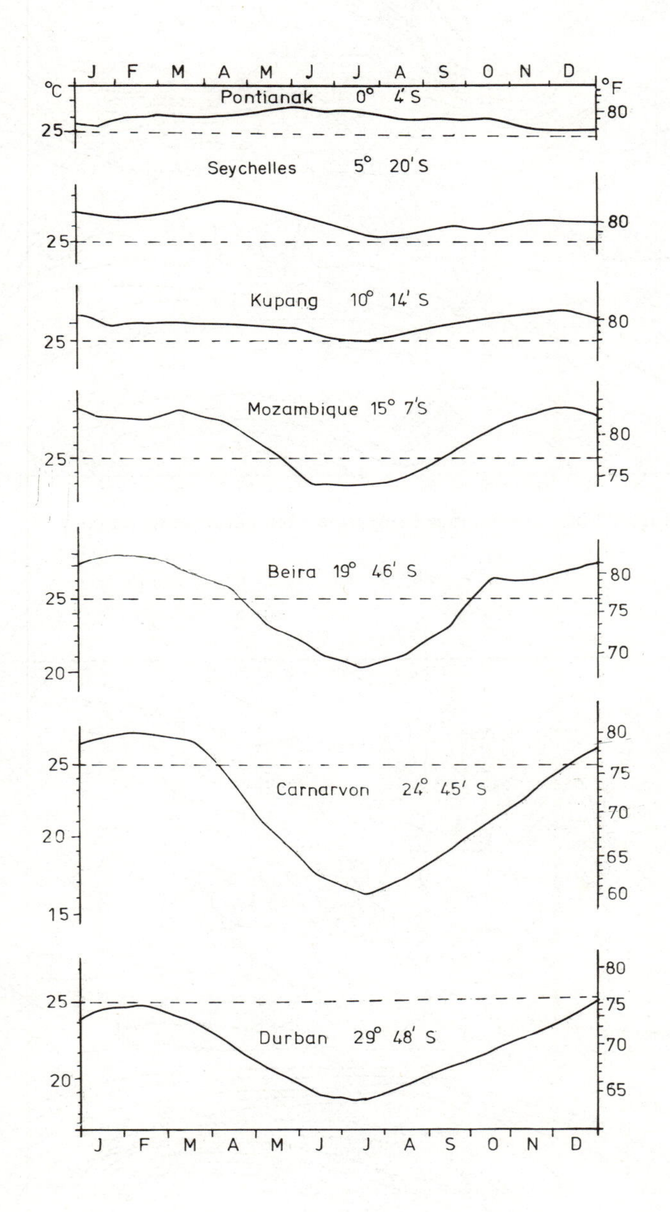
J F M A M J J A S O N D
°C
°F
Pontianak 0° 4'S
25
80
Seychelles 5° 20'S
25
80
Kupang 10° 14' S
25
80
Mozambique 15° 7'S
25
80
75
Beira 19° 46' S
25
20
80
75
70
Carnarvon 24° 45' S
25
20
15
80
75
70
65
60
Durban 29° 48' S
25
20
80
75
70
65
J F M A M J J A S O N D

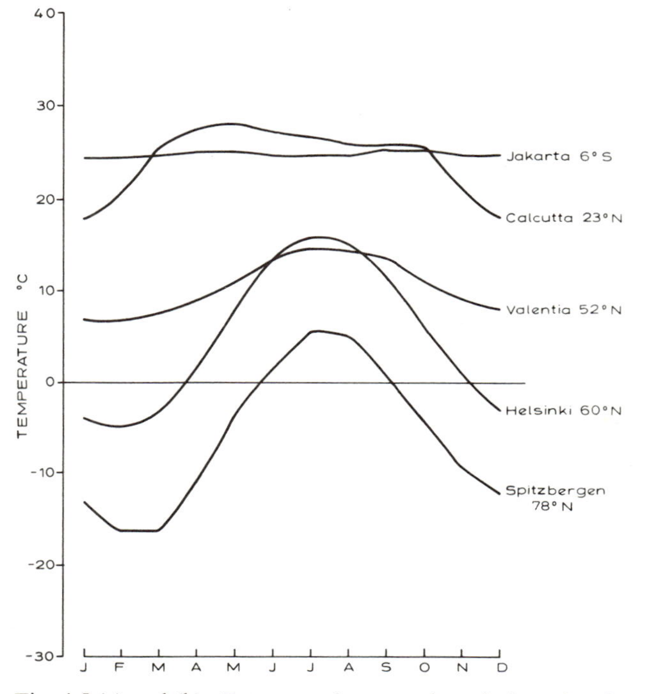

Fig. 4.5 (a) and (b). Patterns of seasonal variations in air temperature in low, middle, and high latitudes (mostly after Petterssen, 1969, and Nieuwolt, 1977)

2 This general equator-to-pole temperature decline is modified by the location of land and water surfaces and the seasonal changes in the sun's position relative to these surfaces.

3 Isotherms are more or less parallel and widely spaced in the southern hemisphere where there is a more nearly homogeneous surface.

4 In the more heterogeneous northern hemisphere isotherms show wide deflections when they pass from ocean to land surfaces.

5 In January isotherms are deflected southward over land and northward over the oceans. Also, within a given latitudinal zone, temperatures are low on land but high on ocean surface.

6 In July, the situation is reversed with isotherms pushed far north over the land surface.

7 Generally speaking, there is greater thermal uniformity with regard to both seasons and places in the tropics than in the temperate region. The thermal uniformity is strongest around the equator and decreases poleward with increasing latitude.

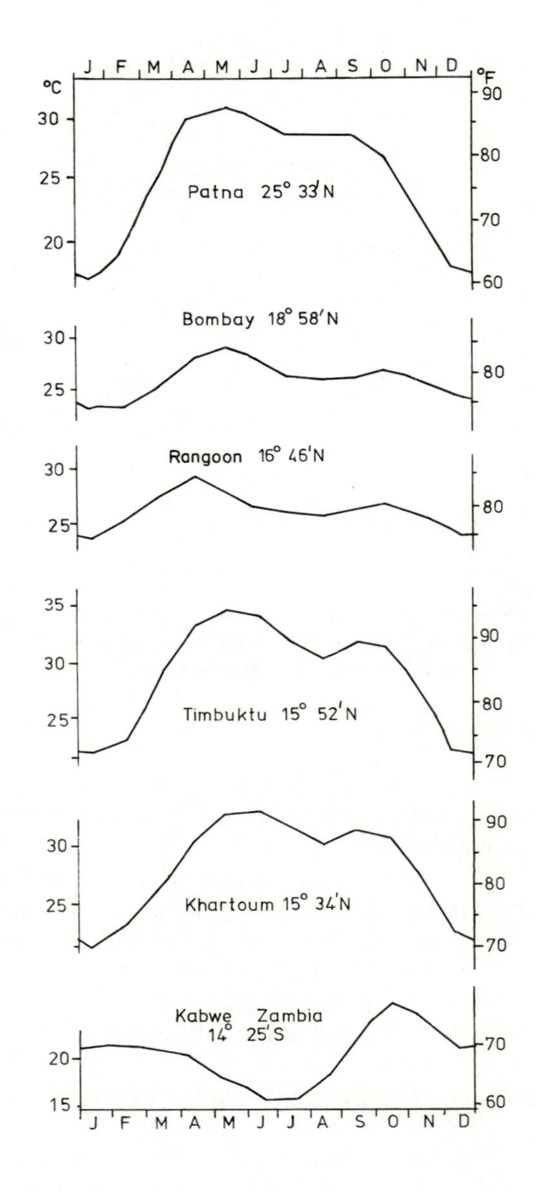
J F M A M J J A S O N D
°C
°F
30
25
20
90
80
70
60
Patna 25° 33′N
Bombay 18° 58′N
30
25
80
Rangoon 16° 46′N
30
25
80
35
30
25
90
80
70
Timbuktu 15° 52′N
30
25
90
80
70
Khartoum 15° 34′N
Kabwe Zambia
14° 25′S
20
15
70
60
J F M A M J J A S O N D

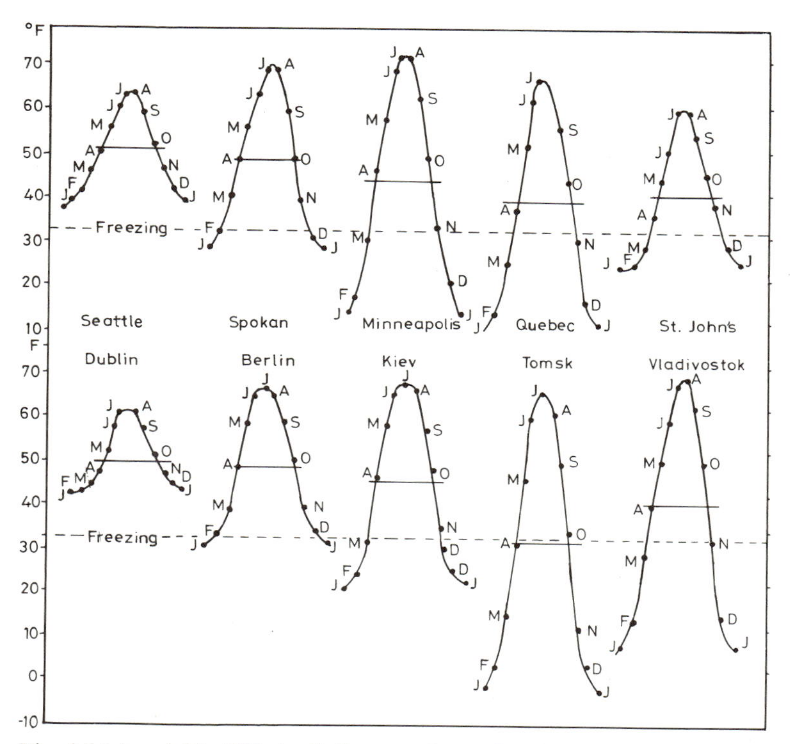

Fig. 4.6 (a) and (b). Effect of distance from the sea (continentality) on seasonal variations in air temperature (after Petterssen, 1969)

Seasonal variations in temperature

The seasonal variations in temperature result largely from the seasonal variations in the amount of insolation received at any given location on the globe. Temperatures are highest in summer when insolation amounts are largest and lowest in winter when insolation receipts are lowest. The seasonal variations in air temperature are greatest in the extratropical areas particularly the continental interiors while they are lowest around the equatorial belt, particularly the water surfaces. We can therefore say that the seasonal variations in temperature increase with latitude and the degree of continentality.

Fig. 4.5 shows examples of seasonal variations in temperatures in the low and high latitudes. In the equatorial zone, the sun is in the zenith twice a year at the equinoxes and temperatures are highest then. The lowest temperatures occur at the solstices. The temperature curve for Jakarta (6 °S 107 °E) is typical of the equatorial zone. Because the sun is high in the sky throughout the year temperature variations are rather small. Human comfort is therefore determined more by humidity than temperature. However, with increasing latitude and degree of continentality, greater variations in the annual march of temperatures occur. With increasing latitude, there are wider variations in the altitude of the sun in the course of the year, particularly between summer and winter seasons. Also, with increasing latitude the days become longer during the summer season while the nights become shorter. During winter the situation is reversed as days become shorter and nights become longer. On the other hand, in the equatorial zone and much of the tropics days and nights are more or less of equal durations virtually throughout the year. Fig. 4.6 shows types of seasonal variations in temperature according to varying degrees of continentality. As would be expected, the annual range is small in maritime locations while it is large in continental locations. This is because the moderating influence of the ocean on temperatures on land decreases with increasing distance towards the hinterland.

Diurnal variations in temperature

The processes that produce seasonality in air temperature values also account for the diurnal variations in temperature, though there are differences in degree. Because the daily cycle is much shorter than the annual cycle we find that the penetration of the surface by solar energy is shallow. Hence the diurnal range of temperature is relatively large. Also, because of the shortness of the cycle, horizontal exchanges of heat are not important except along coasts where there are land and sea breezes (see Chapter 5).

The diurnal range of temperature generally decreases from the equator towards the poles. This is primarily because the daily variation in the elevation of the sun is large in low latitudes and rather small in high latitudes. Also, the diurnal range of temperature is smaller over the ocean than over land partly owing to the already mentioned continentality effect arising from differences in the thermal properties of land and ocean surfaces. Another reason is that the diurnal range of temperature is influenced by cloud cover and the amount of moisture in the air. Clouds reduce insolation during the day and augment the downward radiation from the sky at night. Also, the less the water vapour, the more the amount of outgoing radiation from the earth's surface that can escape to space.

Other factors which influence the diurnal range of temperature of a given surface include wind speed and the conductive capacity of the surface. The range of temperature at the surface is smaller on windy days than on calm days. This is because on windy days the heat exchange affects a deeper layer of air than on calm days. Also, the more the heat conductive capacity of a surface the smaller will be the diurnal range of temperature.

Over the globe as a whole, the diurnal range of temperature is higher over land than over the ocean and is higher in the low latitudes than in the middle and high latitudes. Over the oceans, the diurnal range of temperature is usually less than 0.7 °C and is everywhere smaller than the annual range. Over land, in the low latitudes the diurnal range is higher than the annual range (see Fig. 1.2). In drier areas of this broad zone, the diurnal range is so large that it affects plant and animal life. The large diurnal ranges of temperature in the low latitude deserts are due to the fact that both the ground and air are dry. The conductive capacity of soil decreases with aridity and the moderating influence provided by the formation of dew on the diurnal range of temperatures in the more humid areas is absent.

Latitude and location relative to the oceans influence diurnal ranges of temperature. In the tropics, the day-to-day variations in temperature are not only generally small but the diurnal temperature cycle is also rather regular. In constrast, the diurnal march of temperatures in the temperate region is rather irregular because of the frequent passage of depressions and air masses of strongly contrasting temperatures. The diurnal march of temperatures in the temperate region also varies strongly with the seasons unlike in the tropics. This is because of the variations in the duration of day and night as well as the amount of insolation during the course of the year. In summer, days are much longer than nights while incident radiation is at the highest. On the other hand, winters are characterized by short days, long nights, and low amount of insolation. The diurnal range of temperature is therefore higher in summer than in winter.

With the diurnal cycle of temperature we have corresponding changes in the lapse rate near the ground. The daily range of temperature is highest at the surface of the earth and decreases rather rapidly with height above the ground. Also, in general the lapse rate is rather small in the early morning and becomes steep in the early afternoon as a result of solar heating. Vertical mobility of air which depends on the lapse rate therefore undergoes a diurnal cycle in tune with that of temperature.

Physiological temperature

The temperature experienced by a living organism including man depends on the air temperature as well as the rate of heat loss from that organism. This temperature is called the *physiological temperature* and varies with individuals depending on their characteristics such as general build and weight, type of clothing, physical activities or jobs engaged in, diet, state of health, age, sex, emotional state, and the degree of adjustment to the prevailing climatic conditions.

The heat balance of a human body can be expressed by the equation of the form

$$M \pm R \pm C - E = 0. \tag{4.2}$$

This means that to maintain thermal equilibrium the metabolic heat (M) created chemically within the human body together with the heat gained or lost by radiation (R) and convection (C) and the heat lost by evaporation (E) must add up to zero. The average human body is most efficient at a core temperature of 37 °C. A fall in the body temperature can be prevented by increasing the metabolism rate through voluntary (e.g. work or exercise) or involuntary muscular activity (shivering). It can also be prevented by increasing the amount of heat gained through radiation and convection. To prevent overheating, the human body usually resorts to sweating and panting. Overheating can also be prevented by eliminating heat gained by radiation and convection or increasing the heat lost through evaporation.

Thus, physiological temperature is a function of the ambient thermal environment and the efficiency and speed of evaporation. The ambient thermal environment is determined by the balance between radiation gain and loss. The efficiency and speed of evaporation is controlled by three factors namely the humidity of the air, wind speed, and the degree of exposure to sunshine.

When the air is humid, evaporation of perspiration from the body is limited and a feeling of oppressiveness so common in the humid tropics is created. On the other hand, dry air encourages evaporation of perspiration from the human body, a process which allows a rapid cooling of the skin as latent heat is used in evaporation. For this reason, the human body can endure high air temperatures much better if the humidities are low than if they are high.

Wind speed is also an important factor influencing the rate of evaporation. If the air is calm, the air layer close to the body becomes more or less saturated and little or no further evaporation takes place. If, however, there is considerable air flow, the constant replacement of air around the body ensures that the evaporation process is maintained and not interfered with. Finally, direct exposure to sunshine limits the efficiency of evaporation as a means of cooling the body. This is because the skin will absorb a considerable amount of heat which cannot readily be got rid off.

Indices of physiological temperature are usually based on air temperature and humidity. Wind speed (i.e. ventilation) and exposure to sunshine are generally ignored or held constant. This is because unlike temperature and humidity, they are difficult to control and measure in laboratory experiments. Besides, they can easily be modified in normal living conditions. For instance, electric fans are used in the tropics to alleviate thermal stress while protection against exposure to sunshine during outdoor activities is provided by clothing and shades.

Table 4.2 Comparison of some world-wide comfort zones (modified after Terjung, 1968)

Area	Comfort zone (ET °C)	Investigator
Northern USA	20–22	American Society of Heating and Air Conditioning Engineers (1955)
Southern USA	21–25	American Society of Heating and Air Conditioning Engineers (1955)
Continental Europe	20–26	Mcfarlane (1958)
India	21–26	Malhotra (1955)
Indonesia	20–26	Mom (1947)
Malaya	21–26	Webb (1952)
England	14–19	Bedford (1954)
Northern Nigeria	18–21	Peel (1961)

Of the various indices of physiological temperature the most commonly used is the effective temperature index (ET) which under conditions of light air movement is given by the equation of the form

$$\mathrm{ET} = 0.4(Td + Tw) + 4.8 \tag{4.3}$$

where Td and Tw are the dry-bulb and wet-bulb temperatures measured in °C. The above equation first given by Thom (1959) is sometimes called the discomfort index or the temperature–humidity index. This index has been used in various countries to determine the range of comfort zones for clothed adults

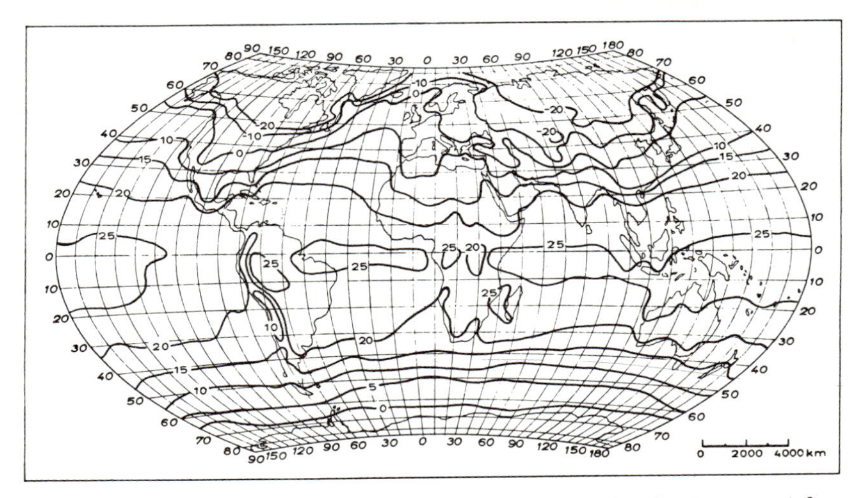

Fig. 4.7 (a). Global pattern of effective temperature index in January (after Gregorczuk and Cena, 1967)

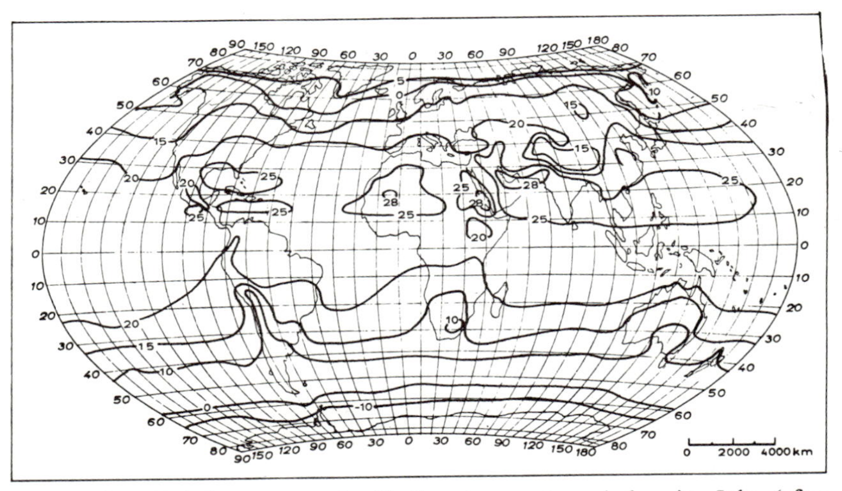

Fig. 4.7 (b). Global pattern of effective temperature index in July (after Gregorczuk and Cena, 1967)

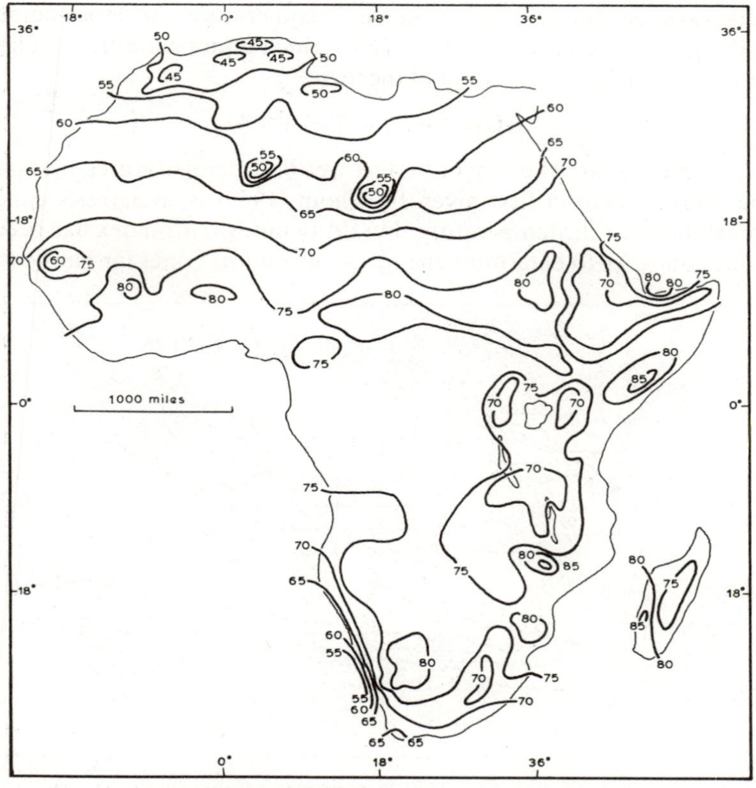

Fig. 4.8 (a). Effective temperature in Africa in January (after Terjung, 1967)

at rest and in the shade with slight air movement (see Table 4.2). Except for England, it appears that there is some kind of agreement as to what the subjects used in the various experimental studies considered comfortable in these diverse climatic zones. We may therefore consider an ET value of 60 °F (18.9 °C) or below as indicating an uncomfortable condition arising from cold stress while an ET value of 78 °F (25.6 °C) or above will indicate heat stress.

The distribution patterns of effective temperature over the earth in January and July are shown in Fig. 4.7. Fig. 4.7(a) shows that the highest ET values in January occur between latitudes 50 °N and 25 °S with ET values of over 25 °C. ET values decrease poleward in both hemispheres but rather more rapidly in the northern hemisphere than in the south. The lowest ET values of − 20 °C occur poleward of latitude 50 °N in North America and about latitude 60 °N in Asia. Poleward of the 60° parallel in the southern hemisphere ET values are about zero but no negative values occur. In July, the highest ET values occur between 10 and 30° north of the equator (Fig. 4.7(b)). ET values decrease pole-

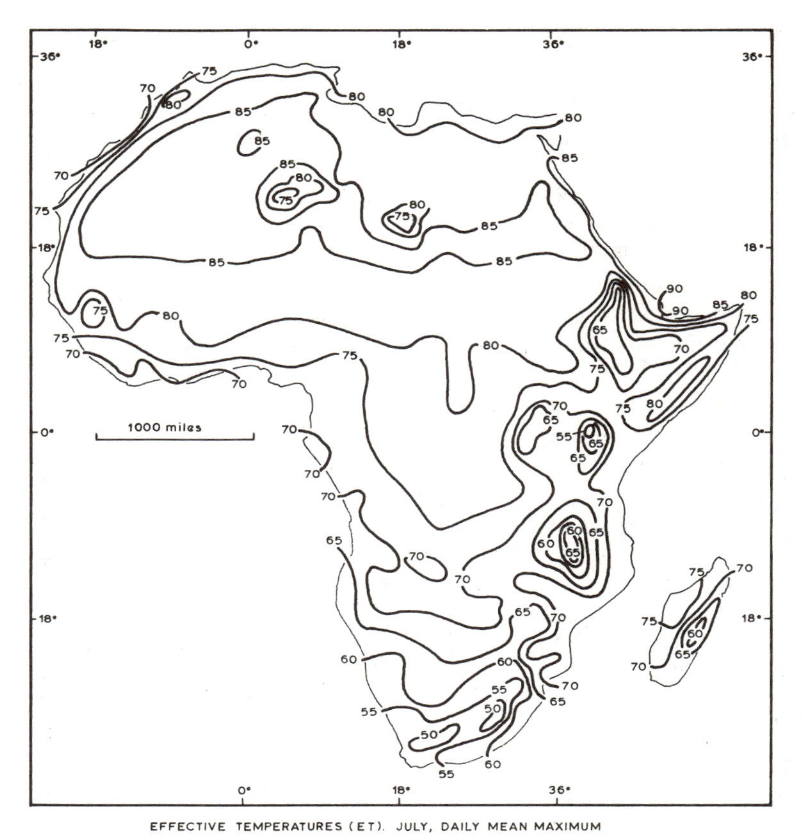

EFFECTIVE TEMPERATURES (ET). JULY, DAILY MEAN MAXIMUM

Fig. 4.8 (b). Effective temperature in Africa in July (after Terjung, 1967)

ward in both hemispheres but more rapidly in the southern hemisphere, where ET values of – 10 °C and below occur poleward of the 60° parallel. In the northern hemisphere, ET values are everywhere above zero. The lowest values of 5 °C and below occur poleward of latitude 70° north.

Thus, the spatial distribution and values of ET are rather similar to those of air temperature. These similarities are modified by the influence of humidity sometimes considerably where air temperatures are high but relative humidities are low. The largest differences between air temperature and effective temperature (ET) therefore occur in desert areas.

More detailed distribution patterns of ET values over the continent of Africa are portrayed in Fig. 4.8. In January during the northern hemisphere winter, ET values range from below 45 °F (7.2 °C) in the north to over 85 °F (29.4 °C) in some areas of southern Africa. Along the coast of south Africa and southwest Africa, ET values are much lower than the latitudinal averages owing to the effect of the cold Benguella current. Over the continent as a whole the weather is relatively cool and physiological heating as perceived by the average person is less widespread than in July.

July represents summer condition in the northern hemisphere and winter condition in the southern hemisphere. ET values are higher in northern Africa than in other areas of Africa.

Relatively low ET values occur in southern Africa and the hilly areas of East Africa. The highest ET values of 85–90 °F (29.4–32.2 °C) occur along the Red Sea Coast in the Somali Republic. This same area has relatively high ET values in January and must rank as one of the most sultry areas of Africa.

In extratropical areas with a well defined cold season the index that provides the most useful assessment of cold discomfort is the wind-chill index of Siple and Passel (1945). This index is a measure of the quantity of heat which the atmosphere can absorb within one hour from an exposed surface that is one metre square. The index was developed from experimental work carried out in the Antarctica and was based on the freezing rate of water sealed in small plastic cylinders under known conditions of temperature and wind.

Using empirical evidence, the wind-chill index was expressed as follows

$$H = (10.45 + 10\sqrt{V} - V)(33 - T) \tag{4.4}$$

where H is the heat loss in kcal $m^{-2}h^{-1}$, V is wind speed in metres per second and T is air temperature in °C. It should be noted that the wind-chill index actually measures the cooling power of wind and temperature in complete shade without regard to evaporation and the cooling rate is based on a neutral skin temperature of 33 °C. A number of different combinations of wind speed and temperature can give the same amount of cooling power. The wind-chill index has also been criticized on the ground that the index represents only the dry convective cooling power of the atmosphere. The index has, however, been found very useful in studies of incidence of frost-bite. Values of clothing thickness required to maintain the body in thermal equilibrium are also closely related to values of the wind-chill index. Terjung (1966) has provided a useful sensation scale in terms of values of the wind-chill index as follows.

Sensation	Values of wind-chill in cal. $m^{-2}s^{-1}$
Exposed flesh freezes	> 400
Bitterly cold	325–400
Very cold	275–325
Cold	225–275
Very cool	160–225
Cool	80–160
Pleasant	50–80
Warm	< 50

References

Gregorczuk, M. and Cena, K. (1967). Distribution of effective temperature over the surface of the earth. *Int. J. Biometeor.* **11**(2), 145–149.

Nelkon, M. and Parker, P. (1970). *Advanced Level Physics* (3rd edn). Heinemann, London.

Petterssen, S. (1969). *Introduction to Meteorology* (3rd edn). McGraw-Hill, New York.

Siple, P. A. and Passel, C. F. (1945). Measurements of dry atmospheric cooling in subfreezing temperatures. *Proc. Am. Phil. Soc.* **89**, 177–199.

Terjung, W. H. (1966). Physiologic climates of the conterminous United States: A bioclimatic classification based on man. *Annals Assoc. Amer. Geogr.* **56**, 141–179.

Terjung, W. H. (1967). The geographical application of some selected Physio-climatic indices to Africa. *Int. J. Biometeor.* **11** (1), 5–19.

Terjung, W. H. (1968). World patterns of the distribution of monthly comfort index. *Intern. J. of Biomet.* **12**, 119–151.

Thom, E. C. (1959). The discomfort index. *Weatherwise*, **12**, 57–60.

W. M. O. (1971). *Guide to meteorological instruments and observing practices.* W. M. O., Geneva.

CHAPTER 5

Atmospheric Circulation

Scales of atmospheric motions

The atmosphere is constantly in motion. Atmospheric motion is the sum of two main components—movement relative to earth's surface (i.e. wind) and movement as a whole with the earth spinning on its axis. The latter movement has important effects on the direction of winds relative to the earth (Atkinson, 1972). There are two dimensions to the motion of the atmosphere relative to the earth's surface—horizontal and vertical dimensions. The motion itself occurs on different time and spatial scales.

The basic underlying cause of atmospheric motion whether in the horizontal or in the vertical is the imbalance in net radiation, moisture, and momentum between the low and high latitudes on the one hand and between the earth's surface and the atmosphere on the other. Other factors which influence atmospheric circulation are topography, distribution of land and water surfaces, and ocean currents.

Earlier in Table 1.1 the three main scales of meteorological motion systems were presented. Emphasis in the table is on weather systems within the atmosphere and the various areal and time scales on which they occur. A similar classification can be made in respect of atmospheric motion or circulation. Thus, we have primary, secondary, and tertiary atmospheric circulations in decreasing order of magnitude of both their areal and time scales.

The primary circulation is the general circulation of the atmosphere described by Barry and Chorley (1976) as the large-scale or global patterns of wind and pressure which persist through the year or recur seasonally. It is the general circulation that really determines the pattern of world climates. For instance, because the general circulation tends to arrange itself in latitudinal zones, world climates also tend to occur in zones. Embedded within the general circulation are the secondary circulation systems such as the middle latitude depressions and anticyclones and the various tropical disturbances. Compared to the general circulation of the atmosphere, these circulation systems are relatively short-lived and move rather rapidly. The tertiary circulation systems consist mainly of local wind systems such as land and sea breezes, lee waves, katabatic and anabatic winds. These circulation systems are highly localized, being controlled largely by local factors and their life spans are considerably shorter than the secondary circulation systems. In this chapter we will consider the features of the general circulation of the atmosphere. The secondary circulation systems are described in the next chapter. But first let us consider the factors which control air movement over the earth's surface. Vertical air motion though important is relatively small compared to horizontal air motion. It is considered in Chapter 7 because of its role in evaporation–condensation processes.

Laws of horizontal motion

There are four major controls on the horizontal movement of air near the earth's surface. These are:

1 the pressure gradient force,
2 the coriolis force,
3 the centripetal acceleration, and
4 the frictional force.

The primary cause of air motion is the development and maintenance of a horizontal pressure gradient which serves as the motivating force for air to move away from areas of high pressure towards areas with lower pressure. Horizontal differences in pressure are created by thermal and/or mechanical causes although these are often not distinguishable. The pressure gradient force is also inversely proportional to air density. The pressure gradient force per unit mass is mathematically expressed as

$$-\frac{1}{\rho}\frac{\mathrm{d}p}{\mathrm{d}n} \tag{5.1}$$

where ρ is air density and $\mathrm{d}p/\mathrm{d}n$ is the horizontal pressure gradient. This means that the closer the isobar spacing the more intense is the pressure gradient and the greater the wind speed.

Once air is forced to move by the pressure gradient force it is immediately affected by the coriolis or deflective force due to the earth's rotation. Coriolis force was first described mathematically by the French scientist Gaspard de Coriolis in 1835 and demonstrated by American meteorologist William Ferrel in 1856. Because of the rotation of the earth, there is an apparent deflection of moving objects including air to the right of their line of motion in the northern hemisphere and to the left in the southern hemisphere as viewed by any observer on the earth's surface. This deflective force per unit mass is mathematically expressed as

$$-2wV\sin\theta \tag{5.2}$$

where w is the angular velocity of spin of the earth on its axis (about 15° per hour or 7.29×10^{-5} radians/s), V is the velocity of the mass and θ is the latitude. Thus the magnitude of deflection is proportional to the velocity of mass and the sine of latitude. For a given velocity, the coriolis effect is highest at the poles and decreases with the sine of the latitude, becoming zero at the equator. Coriolis force always acts at right angles to the direction of air motion to the right in the northern hemisphere and to the left in the southern hemisphere (Barry and Chorley, 1976).

If a body during motion follows a curved path there must be an inward acceleration towards the centre of rotation. This centripetal acceleration is mathematically expressed as

$$-\frac{mV^2}{r} \tag{5.3}$$

where m is the moving mass, V is its velocity and r is the radius of curvature. The centripetal acceleration may also be regarded as a centrifugal force operating radially outward. Such a force is equal in magnitude but opposite in sign to the centripetal acceleration. The magnitude of centripetal acceleration

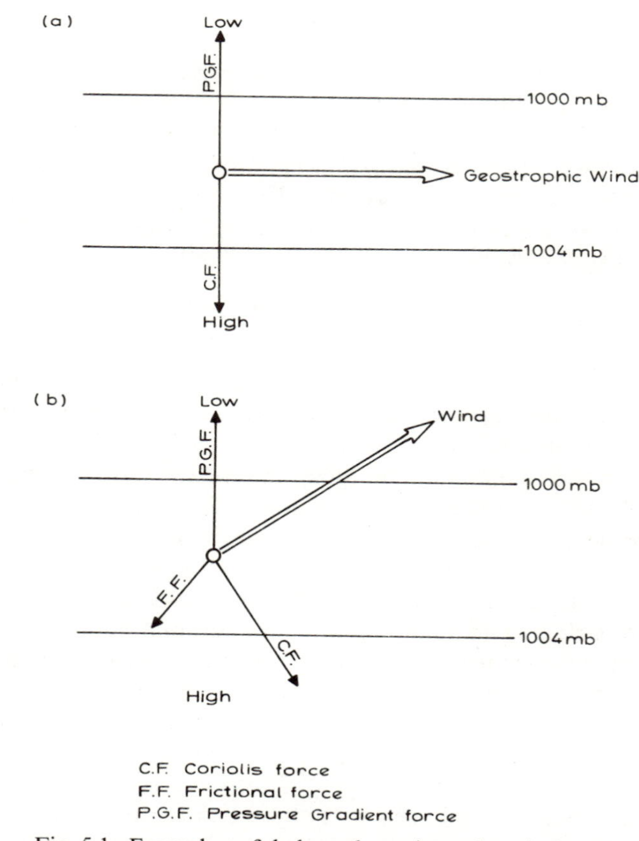

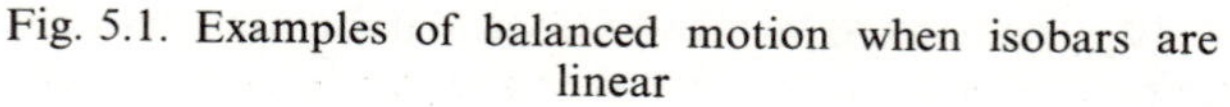

Fig. 5.1. Examples of balanced motion when isobars are linear

is small so that it only becomes important where high velocity winds move in very curved paths as in an intense low pressure system.

Finally, close to the earth's surface a fourth force—that of frictional force—helps to control the speed and direction of horizontal air motion. The frictional force is due to the drag by the earth's surface on air motion. Frictional force acts against the wind and reduces its speed. This also causes a decrease in the coriolis force which is partly dependent on velocity as mentioned earlier.

All the forces described above do not necessarily operate to control wind direction and speed at a given time or given place. The examples of the balance of forces in respect of linear isobars are shown in Fig 5.1. In Fig. 5.1(a) the wind blows parallel to the isobars, i.e. more or less at right angles to the pressure gradient. This is because the pressure gradient force is exactly balanced by the coriolis force acting in the opposite direction. Such a wind is known as the *geostrophic wind* and can be observed in the free atmosphere where friction is non-existent. The case shown in Fig. 5.1(a) is the northern hemisphere. In the southern hemisphere, the high pressure core will be on the left and the low pressure on the right when viewed downwind.

The velocity of the geostrophic wind (Vg) is given by the formula

$$Vg = \frac{1}{2w \sin \theta \rho} \frac{dp}{dn} \tag{5.4}$$

where w, θ, ρ and dp/dn are as defined in equations 5.1 and 5.2. This indicates that velocity is inversely related to latitude. Except in the low latitudes where Coriolis deflection approaches zero, the geostrophic wind is a close approximation to the observed motion in the free atmosphere.

From the earth's surface to about 500–1000 metres the frictional force is operative and the wind blows across the isobars in the direction of the pressure gradient (see Fig. 5.1(b)). The angle at which the wind blows across the isobars increases with increasing frictional drag created by the earth's surface. It is about 10–20° at the surface over the sea and 25–35° over land. With increasing height above the surface whether on land or over the sea, the frictional drag decreases. A kind of Ekman spiral of wind with height therefore occurs if we consider the theoretical profile of wind velocity with height under conditions of mechanical turbulence (see Fig. 5.2).

The patterns of air flow arising from the balance of forces in low and high pressure systems in the northern hemisphere are shown in Fig. 5.3. In a low pressure system balanced flow is maintained in a curved path by the excess of the pressure gradient force over the coriolis force giving the net centripetal acceleration inward. This wind is known as the *gradient wind*. In the case of the high pressure system the inward acceleration is due to the excess of coriolis force over the pressure gradient force. If the pressure gradients in both systems are assumed to be equal, then the wind velocity around the low pressure must be less than the geostrophic value (subgeostrophic) while that around the high pressure is supergeostrophic, i.e. more than the geostrophic value. This effect is, however, masked by the fact that the pressure gradient in a high is

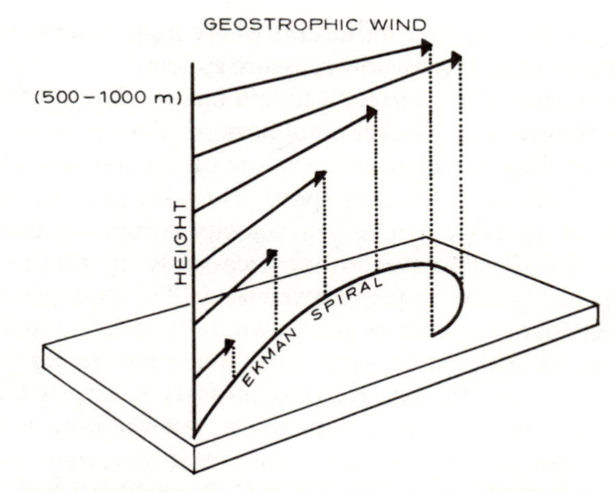

Fig. 5.2. The Ekman spiral of wind with height in the northern hemisphere (reproduced by permission of Methuen from Barry and Chorley, 1976)

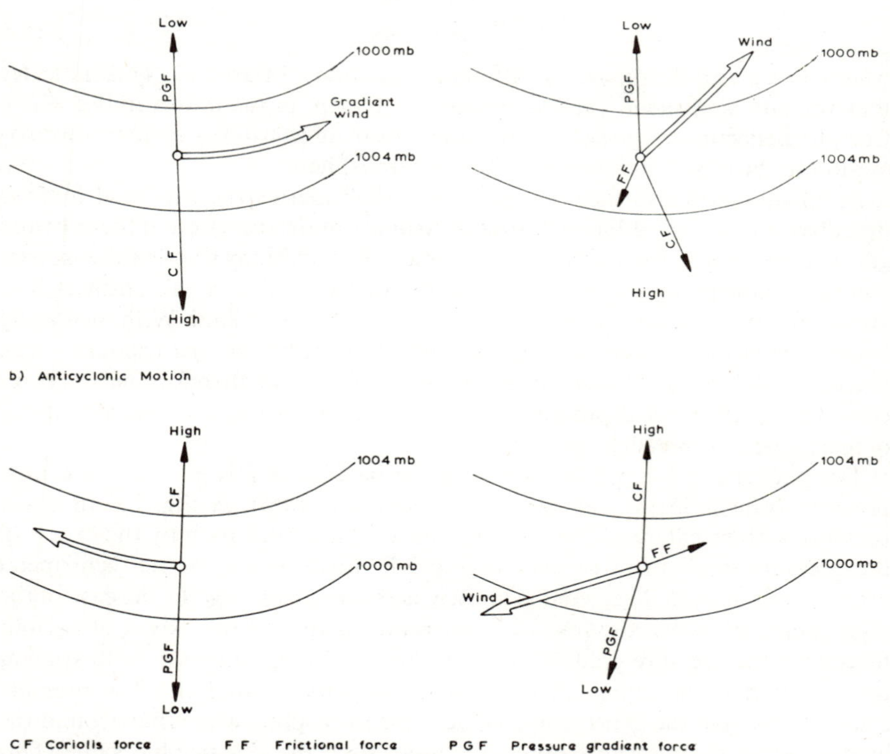

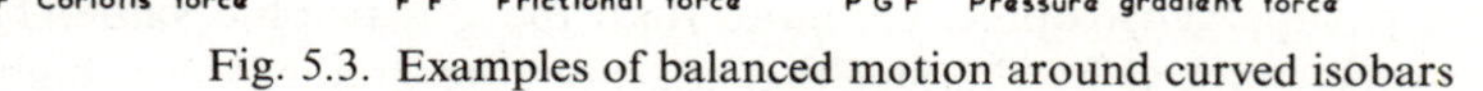

Fig. 5.3. Examples of balanced motion around curved isobars

usually much weaker than that in a low. In both the low and high pressure systems, the effect of the frictional force is to make the winds blow at an angle across the isobars as well as decrease their velocities.

Large-scale vertical motion in the atmosphere

Vertical motion within the atmosphere occurs on two major scales—large scale and small scale. The large-scale vertical motion takes place over large areas of several thousand square kilometres on a time scale of a few metres per second. The small-scale vertical motion occurs over small areas of a few hundred square kilometres with a time scale of 1–30 metres per second. The two types of vertical motions are different because they are caused by different mechanisms. Small-scale vertical motions are generally induced directly by the earth's surface and their continuation is largely dependent on the thermal and humidity structure of the overlying atmosphere (Atkinson, 1972). The motion is induced either mechanically or thermally or by a combination of both processes. Since these motions give rise to condensation in the atmosphere they are considered in Chapter 7.

On the other hand, mass uplift or descent of air occurs primarily in response to dynamic factors which are related to horizontal flow. Large-scale vertical motion of air is only secondarily affected by air mass stability. Wind speed often varies from the geostrophic value owing to local accelerations or decelerations in horizontal flow. If in a unit volume of air, more air leaves than arrives due to local acceleration, there is loss of mass in that volume. This is known as *divergence*. If on the other hand, there is deceleration in the horizontal flow, air will pile up in the volume and *convergence* will occur. Since the atmosphere is a continuous medium, configurations of divergence and convergence are linked.

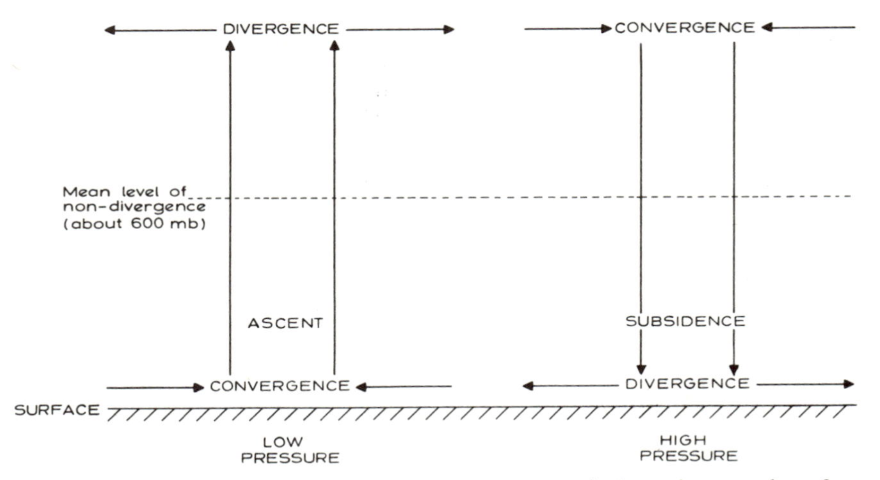

Fig. 5.4. Relationship between divergent patterns, vertical motions, and surface pressure

As shown in Fig. 5.4, if divergence overlies convergence there is uplift; but if convergence overlies divergence, subsidence occurs. Thus, if all winds were geostrophic, there would be no convergence or divergence and no large-scale vertical motion and hence no weather.

Convergence and divergence can also be induced by large-scale topographic features such as the Rockies or Andes or even the land/ocean boundary. The latter is known as coastal convergence or divergence. But convergence and divergence are usually dynamic phenomena of the free atmosphere well beyond the boundary layer of the atmosphere.

Major features of the general circulation of the atmosphere

As pointed out earlier, the large-scale motion of the atmosphere in time and space is what we call the general circulation of the atmosphere. This circulation has both vertical and horizontal components and is controlled by several factors as noted earlier. Basically, the general circulation of the atmosphere is driven by the imbalance in radiation, moisture, momentum, and mass between the low and high latitudes. The global patterns of radiation which were considered earlier on in Chapter 3 show that there is surplus energy in the low latitudes and deficiency elsewhere. In Chapter 7, the global pattern of moisture will be considered. Suffice it to say at this stage, that there is surplus of moisture in the low latitudes which is transported towards the higher latitude to make up for the deficiency there.

Now let us consider the earth's angular momentum, generally recognized as the second most important factor controlling the general circulation of the atmosphere. The atmosphere not only rotates with the earth but also moves on its own around the earth's axis. The atmosphere thus possesses angular momentum. The angular momentum (also called moment of momentum) per unit mass of a body rotating about a fixed axis is proportional to its velocity and its distance from the axis of rotation. With a uniformly rotating earth and atmosphere there is conservation of angular momentum, i.e. the total angular momentum remains constant. This means that if a mass of air changes its position on the earth's surface such that its distance from the axis of rotation is affected its angular velocity must change for angular momentum to remain constant. Angular momentum is highest at the equator and decreases polewards to become zero at the pole, the axis of rotation itself. Hence, an air parcel moving poleward acquires progressively higher eastward velocities.

The low latitudes are generally dominated by easterly winds, the trades, while the middle latitudes are dominated by the westerlies. The friction between the easterlies and the earth rotating from west to east generates easterly angular momentum in the low latitudes. This excess momentum is transferred to the sink areas in the middle latitudes where the westerlies continually impart westerly momentum to the earth by friction. Were momentum not continually replenished from the low latitudes, the westerlies would die off within 10 days due to frictional dissipation of energy.

Angular momentum is transferred from low to high latitudes in the following ways (Chandler, 1967):

1 by movements in the 'Hadley Cell' in the low latitudes;
2 by travelling atmospheric disturbances particularly in the upper troposphere of subtropical and anticyclonic belt;
3 by the high level tropospheric pressure waves and their accompanying family of surface cyclonic and anticyclonic disturbances.

We may now consider the major features of the general circulation of the atmosphere.

The patterns of pressure and global wind systems near the earth's surface are shown in Fig. 5.5. The effects of differential heating of land and water surfaces are largely neglected, but the coriolis effect has been taken into consideration so that the winds shown on the diagram are deflected to the right of their path in the northern hemisphere and to the left of their path in the southern hemisphere. There are low pressure belts around the equator and around latitudes 60° north and south of the equator. High pressure belts occur around the poles and around latitudes 30° north and south of the equator. The low pressure belt

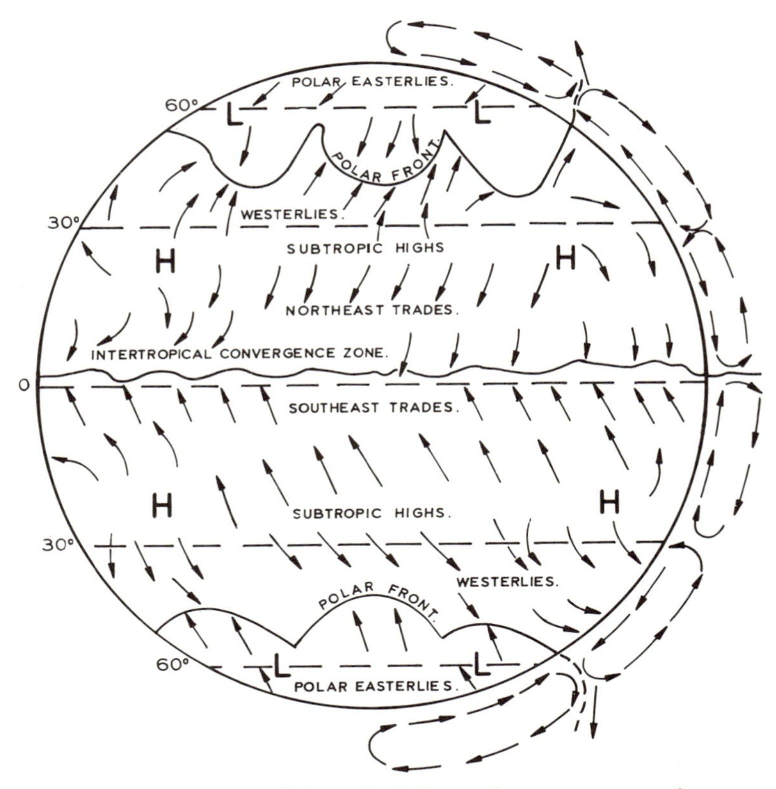

Fig. 5.5. Pressure belts and winds on a homogeneous earth

around the equator is essentially thermal in origin, i.e. due to solar heating. The subpolar low pressure belts around latitudes 60 °N and S of the equator are essentially dynamic in origin. They are caused by the rotation of the earth which causes a polar whirl and therefore a tendency towards low pressure around the poles. Because of the intense cold around the poles, however, the dynamic effect is masked by the thermal effect. Pressure is therefore high around the poles because of the intensely cold air prevailing there.

The subtropical highs have been explained as due to the effects of one or more of the following mechanisms:

1 the piling up of poleward moving air as it is increasingly deflected eastwards through the earth's rotation and the conservation of momentum;
2 the sinking of poleward currents aloft by radiation cooling; and
3 the necessity for a high pressure belt near 30° parallel to separate approximately equal zones of observed easterly and westerly winds.

In response to the above pressure distribution patterns there are six wind systems in all, three in each hemisphere. In the northern hemisphere are the northeast trade winds, the westerlies, and the polar easterlies while in the southern hemisphere we have the southeast trades, the westerlies, and the polar easterlies. It is important to note that the southeast trades cross the equator to become southwesterly winds as in the West African region while the northeasterlies become the northern westerlies in the southern hemisphere after crossing the equator.

Apart from the surface winds shown in Fig. 5.5, other notable principal features of the circulation of the atmosphere include the following (see Barry, 1967):

1 large amplitude unstable disturbances in the westerlies in the middle and high latitudes;
2 slow steady easterly flow in the low latitudes including some intense vortices (see Chapter 6);
3 strong narrow currents of air called the *jet streams* in the upper troposphere over the middle latitudes.

The long waves in the westerlies were first studied and described by an American Meteorologist, Rossby. Hence, the waves are often called Rossby waves. These waves occur in the middle or high troposphere within the upper westerlies. The waves travel more slowly than the winds blowing through them. Sometimes they remain stationary and may even retreat westward. Four or five long waves typically occur on a hemispherical chart with wavelength of the order of 2000 km. Rossby waves are partly thermal in origin and are partly the consequence of the effect of mountain barriers on air flow since the troughs do not appear to vary seasonally in their locations. Rossby waves have been shown to have connections with mid-latitude weather as they exert influence on the positions of surface depressions and anticyclones. Depressions form or deepen in an area of upper divergence like under the right limb of an upper trough. Conversely, anticyclones form or intensify in an area of upper convergence like the left limb of an upper trough.

The jet stream is a ribbon of air some thousands of kilometres in length, hundreds of kilometres wide and a few kilometres deep with a minimum wind speed of about 120 km per hour. Two main types of jet stream are recognized: the subtropical and polar front jet streams both of which are found just below the tropopause. The subtropical jet stream is thought to be dynamic in origin being a product of the rotation of the earth. The atmosphere has its greatest angular momentum at the equator. The rising air spreading out northwards and southwards from the equator moves faster than the latitude to which it is blowing. The air is deflected to the right in the northern hemisphere and to the left in the southern hemisphere and becomes concentrated as the subtropical jet stream around latitude 30° (Riley and Spolton, 1974). The subtropical jet stream is relatively constant in position in a given season.

In contrast the polar front jet stream is highly variable in position from day to day over a wide range of temperate latitudes. The polar front jet stream is produced by a temperature difference and is closely related to the polar front. Hence, it is of meteorological importance. The subtropical jet stream is also thought to play some role in the development of Asiatic monsoon circulation. Jet streams are of great importance in aviation. A plane that moves with the jet stream will conserve fuel and has its speed increased while one that moves against a jet stream will be slowed down and burn more fuel. This was in fact how the existence of jet streams was discovered during World War II bombing missions in the Far East.

A third jet stream—the polar night jet stream—is found in the stratosphere in the high latitudes in winter.

Models of the general circulation of the atmosphere

Thus, the general circulation is really very complex owing to the various factors involved. There are complexities arising from:

1 differences in the thermal properties of the earth's surface particularly between land and water surfaces;
2 variations in the earth's topography;
3 transformations of energy from one form to another within the atmosphere; and
4 the different and interacting scales of motions.

These complexities and the problems posed by inadequate observational data on the atmosphere both in the horizontal and in the vertical have so far precluded the development of a satisfactory model of the general circulation of the atmosphere.

Various conceptual models of the general circulation of the atmosphere have been presented at various times by different people (see Fig. 5.6), but our understanding of the atmosphere is still incomplete. Perhaps the first model was

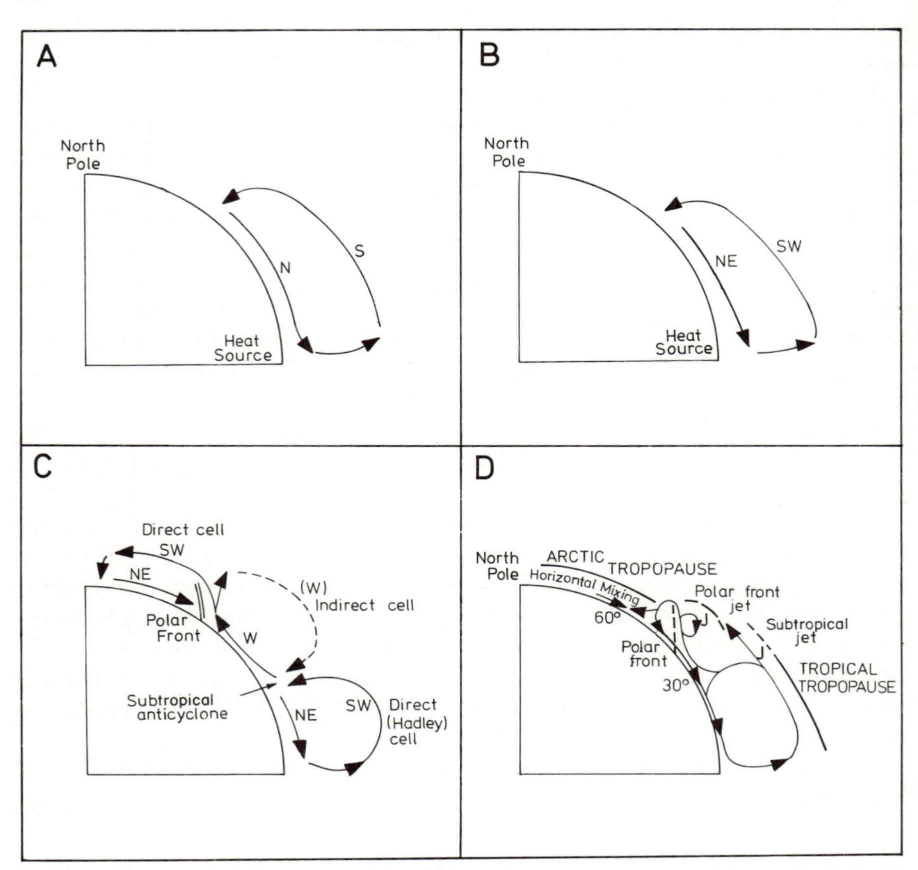

Fig. 5.6. Conceptual models of the general circulation of the atmosphere: A, thermally direct cell on a stationary earth (Halley model, 1686); B, thermally direct cell on a rotating earth (Hadley model, 1735); C, three cell model of the mean circulation (Rossby, 1941); D, the mean meridional circulation in winter (Palmen, 1951)

that presented by Edmund Halley who in 1686 outlined a thermal circulation model with maximum heating in low latitudes and a thermally direct cell accounting for the equatorward flow of the trade winds. This model was slightly improved upon by George Hadley who in 1735 incorporated the effects of the earth's rotation to explain the northeasterly and southeasterly trades and envisaged a compensatory southwesterly counter current above the trades. This meridional-plane circulation is still valid for the tropics and is now referred to as the 'Hadley cell'.

It was not until 1856 that Ferrel postulated a much better model than the earlier two. For the first time the westerly wind belts, overlooked by the earlier two models, were brought in. The three-meridional cells model of Ferrel was again put forward by Bergeron in 1928 and Rossby in 1941 with minor amendments. The Palmen model of 1951 is a further modification of the three-cell model as originally put forward by Ferrel in 1856.

Because of the complexity of the general circulation and the need to isolate the fundamental processes at work mathematical and experimental methods are now used. Important contributions have been made by laboratory studies of rotating fluids to the theory of the general circulation of the atmosphere. The development of mathematical models of the general circulation has received great impetus from advances in space and computer technology. Data are now being obtained from inhospitable and inaccessible land areas of the world as well as the oceans, thanks to weather satellites. Such experimental and mathematical models have further improved our understanding of the general circulation of the atmosphere.

The tropics is known to play a vital role in the general circulation of the atmosphere. All the things the general circulation of the atmosphere is trying to redistribute—energy, moisture, momentum—are in excess supply in low latitudes and in deficiency in the middle and high latitudes. Recent observations from satellites indicate that the energy source in the tropics is even larger than previously estimated. As mentioned earlier, the tropics help to sustain the kinetic energy of the westerlies which would die off within 10 days were it not for the import of momentum from the tropics. As Riehl (1969) has aptly noted the tropical circulation is not driven from the higher latitudes. Rather, the tropical circulation helps to drive the circulation in the middle and upper latitudes. The vigour of the westerlies thus provides an indirect measure of the general circulation of the atmosphere itself.

The overall strength of the westerlies can be assessed by measuring the average hemispheric pressure gradient between 35° and 55° parallels and converting this into geostrophic west wind. This is known as the *zonal index*. The value of the zonal index is high in winter, when the temperature contrasts between the tropics and the middle latitudes are greatest, and low in summer when the thermal contrasts are least. When the average pressure difference between 35 and 55 °N is more than 8 mb, the zonal index is said to be high, when the pressure difference is less than 3 mb, the index is said to be low.

Certain synoptic patterns are associated with high and low zonal index. The following features are associated with high index:

1 rapid eastward movement of depressions;
2 little meridional exchange of air masses even though the meridional temperature gradient is strong;
3 intense development of the subtropical anticyclones and the Icelandic lows;
4 the Icelandic lows tend to be located eastward of their normal position.

With low zonal index, the following features are observed:

1 the pattern of circulation is strongly cellular;
2 longitudinal temperature gradient is strong;

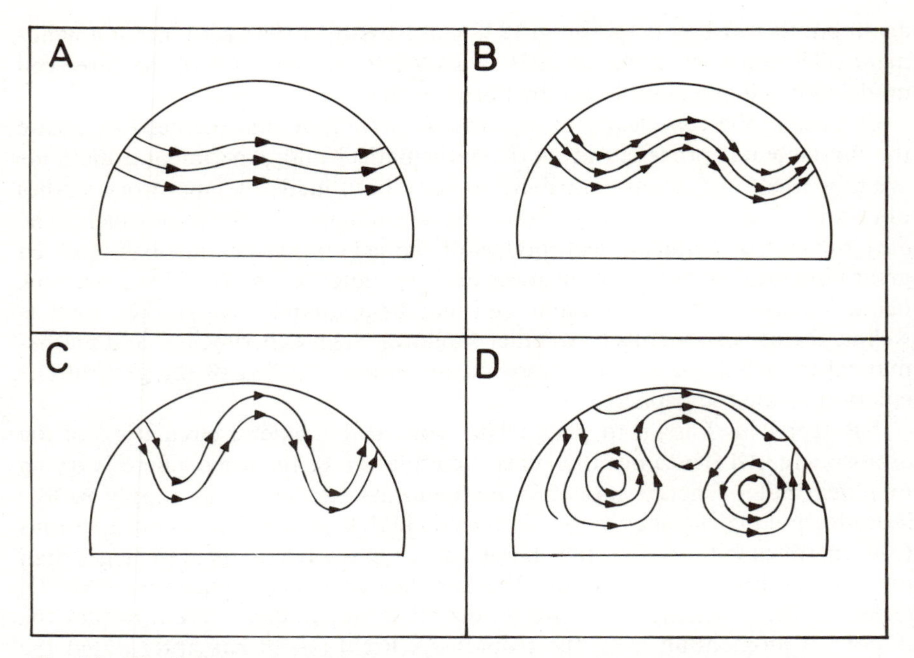

Fig. 5.7. The index cycle: four stages in the changeover from zonal flow (A) to a meridional cellular pattern (D)

3 the Aleutian and Icelandic lows tend to be located west of their normal positions;
4 the development of blocking anticyclones (see Chapter 6).

It has also been established that there are cyclical variations in the zonal index. Over a period of four to six weeks 'index cycles' occur. Fig. 5.7 shows the four stages recognized by Namias in the changeover from the zonal flow of high zonal index to the meridional cellular circulation pattern of low zonal index. In the first stage, the zonal index is high. The jet stream and the westerlies lie north of their mean positions. The westerlies are strong and pressure systems have a dominantly east–west orientation. There is little north–south exchange of air mass. In the second and third stages, the jet stream expands and increases in velocity, undulating with increasingly larger oscillations. In the fourth stage characterized by low zonal index, there is complete break-up and cellular fragmentation of the zonal westerlies which are weak at this time.

Seasonal variations in atmospheric circulation

The mean pattern of atmospheric circulation described above is subject to several important variations. Some of these variations, however, occur regularly in seasonal and/or diurnal cycles. Such variations are of climatological significance, particularly in the tropics which in contrast to the higher latitudes

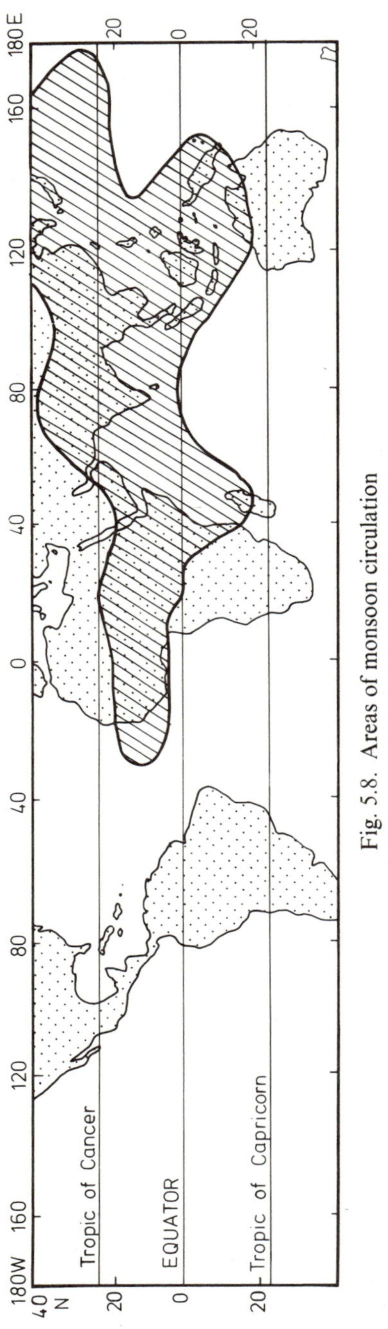

Fig. 5.8. Areas of monsoon circulation

have relatively few disturbances and moving weather systems. Seasonal changes in the circulation of the tropical atmosphere are very small over the large oceans but large over the continents and the adjacent seas. This is because the continents by virtue of their different thermal characteristics (see Chapter 3) create much larger seasonal temperature variations than the oceans. In summer the continents are transformed into centres of low pressure while in winter they are relatively cool compared to the warm oceans. The tropical continents and the surrounding oceans therefore experience a seasonal reversal of wind direction known as the 'monsoons'. The basic and essential cause of the monsoon is the differential heating of large land and ocean areas, changing with the season. The monsoons have consequently been regarded by some people as land and sea breezes on continental and seasonal scales.

Fig. 5.8 shows that areas with monsoon circulations are to be found in the low latitudes with the notable exception of tropical South American where monsoon circulation is poorly developed. The poor development of the monsoon circulation in South America is due primarily to the small size of the continent and its very limited extent to the northern hemisphere. For these reasons strong and persistent thermal lows do not develop over the continent in summer. Also, because the western coast of the continent is influenced by cold current the continent is relatively warmer than the ocean throughout the year thus preventing the formation of high pressure cells over the continent in winter. Although the monsoon circulations in the various parts of the tropics have common major features, regional variations also exist. These variations are brought about by the nature of the earth's surface like the form, size, and relief of the continents as well as by conditions in the upper atmosphere.

The monsoon circulation is best developed in east and southeast Asia for two reasons. The first is the large size of the Asian continent, the largest in the world. The second is the effect of the Tibetan Plateau on air flow. The Tibetan Plateau is an extensive area of high ground elongated in a west–east direction thereby forming an effective barrier between the tropical and the polar air masses (Nieuwolt, 1977). During northern hemisphere winter, radiation losses from the largely snow covered surface of the northern parts of the Asian Continent result in the development of an intense high pressure belt there. Winds move southward and southeastward from this zone of high pressure. On crossing the Equator, these winds recurve into the westerlies over Indonesia. While most of southern and eastern Asia is dominated by the winter monsoon northern India is rarely affected owing to the barrier effect of the Himalayas which prevents the polar air from moving over the Ganges lowlands. Instead, northern India experiences mostly westerly winds. Southern India, however, experiences the winter monsoon. But this monsoon is rather weak and its air masses are not of polar origin. Except over Indonesia and northern Australia where the modified monsoon winds are humid, warm, and unstable, the Asiatic winter monsoons are generally dry, cold, and stable and they bring cold and rather dry winters to most parts of Asia.

In summer, the Asiatic monsoon circulation is more complex but this

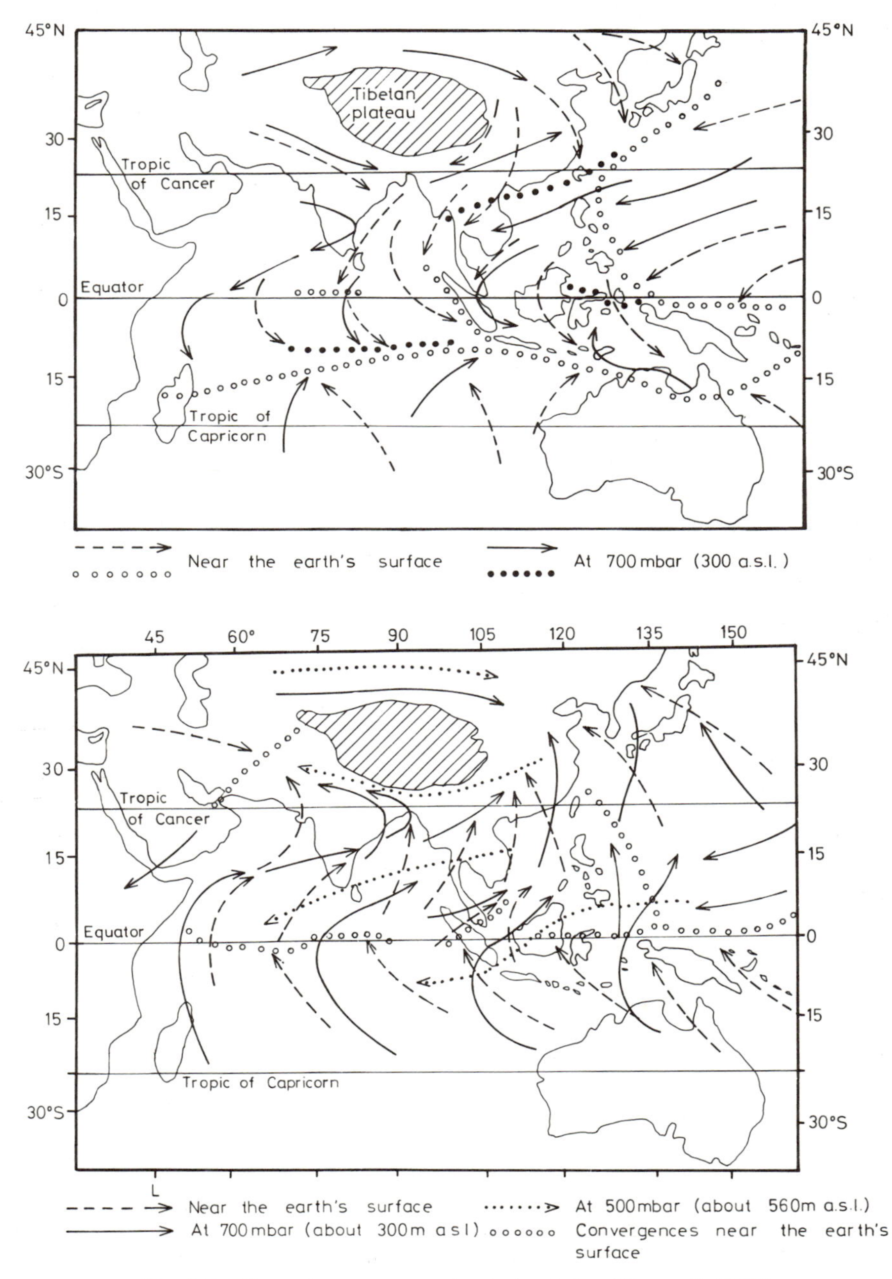

Fig. 5.9. The Asian monsoon circulation (after Nieuwolt, 1977)

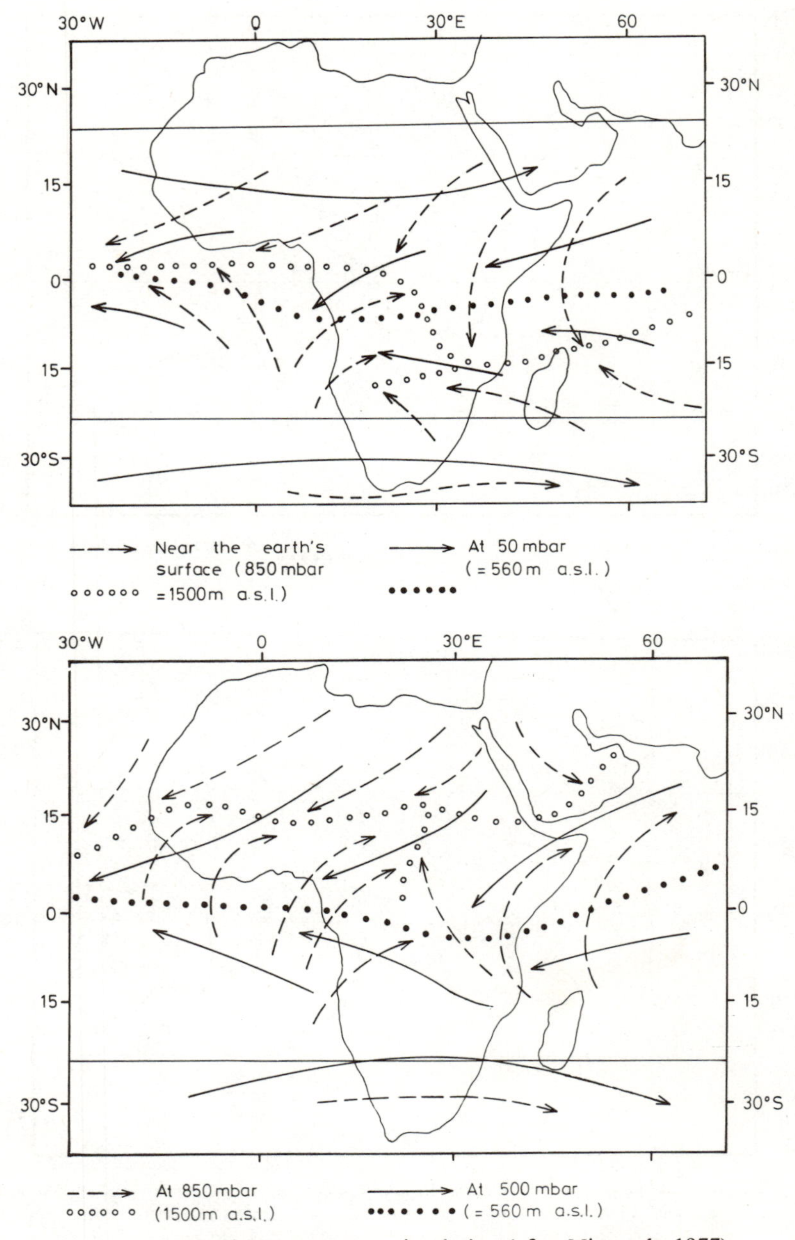

Fig. 5.10. The African monsoon circulation (after Nieuwolt, 1977)

circulation is of great importance as it is the major source of precipitation in many parts of Asia. Series of thermal lows develop over the continent of Asia and the monsoon winds blow towards the land in response. The monsoon develops first over southern China and progresses to Burma and starts over India more than a month later. This delayed start of the monsoon over India has been linked with the upper air circulation about 6000–8000 metres above the earth's surface (Nieuwolt, 1977). All the air masses constituting the summer Asian monsoon are humid and unstable because of their long journey over warm tropical oceans; and they give large amounts of precipitation.

The Australian monsoon circulation can be regarded as an extension of the Asian Monsoon with the seasonal characteristics reversed. During the southern hemisphere winter southeasterly winds blow from the high pressure cell over the southern Pacific. These winds bring dry weather to northern Australia. During southern hemisphere summer, the Asian winter monsoon reaches northern Australia as the northwest monsoon which brings precipitation and warm and humid weather.

The monsoon circulations over Africa are smaller than the Asian monsoon in magnitude in terms of both areal coverage and thickness of the atmosphere involved. The African monsoons are entirely surface winds as they rarely reach levels higher than 5000 metres. In West Africa there are notable differences between the two monsoon winds but in East Africa the two monsoon winds are very similar in their thermal and humidity characteristics and are only differentiated by their directions. This regional variation in the African monsoons is due to the form of the African continent. In the west, there is a large land area north of the equator contrasting with the ocean to the south. In the east, on the other hand, the continent stretches on both sides of the equator, is more mountainous and experiences some influence of the Asian monsoon.

In West Africa, rain is obtained from the summer monsoon winds, the southwesterly winds coming from the Atlantic and blowing towards the thermal low created by solar heating in the interior of the region. The winter monsoon is the dry, stable northeasterly winds from the Sahara which on account of radiational cooling has a high pressure cell located there. The situation in East Africa is a bit more complex for reasons given earlier (see Figs. 5.9 and 5.10).

Diurnal variations in atmospheric circulation

Diurnal wind systems occur frequently and regularly in many tropical areas. They also occur in other areas but rather irregularly and less frequently. This is because such diurnal wind systems tend to be masked by other weather systems, but in the tropics the diurnal wind systems assume some importance in the absence of fronts and strong depressions so common in the temperate region.

There are two major types of diurnal wind systems—land and sea breezes which occur along the coast or near large lakes or other large water bodies; and mountain (katabatic) and valley (anabatic) winds which occur in areas of varied relief.

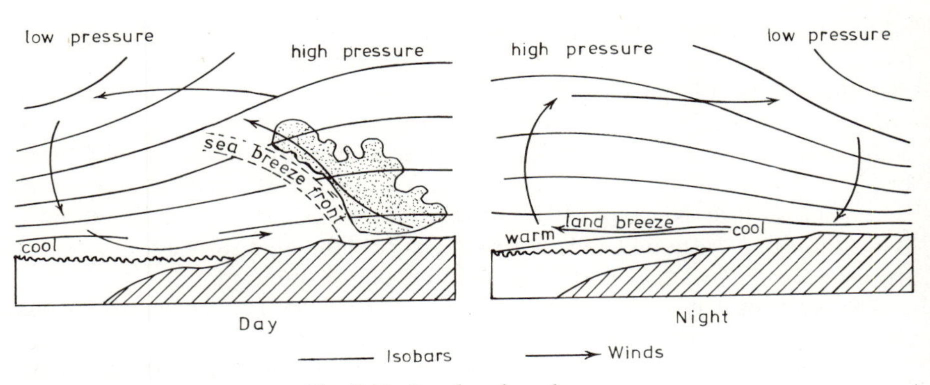

Fig. 5.11. Land and sea breezes

Land and sea breezes are not limited to the tropics but they are strongest and occur most regularly there. They are caused by the thermal differences between the land and water surface. During the day the land heats up more quickly than the water surface for reasons given earlier in Chapter 3. A local thermal low develops over the land with winds blowing from the sea towards the land. This is the sea or lake breeze (see Fig. 5.11). At night the land cools off rapidly while the sea is still warm, the pressure gradient is thus reversed and wind now blows from the land towards the sea. This is the land breeze. The sea breeze is usually stronger than the land breeze and the effect is sometimes felt as far as 60 km inland. The sea breeze starts a few hours after sunrise and is most intense during the early afternoon. The sea breeze is strongest when insolation is most intense. Sea breeze is therefore best developed during the dry season in the tropics and in summer in the temperate region. The sea breeze rarely brings rain but it brings welcome relief from the oppresive heat during the day in many tropical areas. Fishermen also use the land breezes to move out to sea in the early morning and return to land with the sea breezes in late afternoon.

Mountain and valley winds can develop anywhere there are large variations in relief but the winds are particularly strong and regular in the tropics in the absence of fronts and strong depressions. Also, mountain and valley winds are partly thermal in origin. During the day when insolation is intense the more exposed hill slopes are heated more than the valley bottoms. A relatively weak pressure gradient consequently develops with light winds moving upslope (see Fig. 5.12). These are valley or anabatic (upslope) winds. These winds rise above the ridge line and feed an upper return current to compensate for the valley winds. Anabatic winds are often accompanied by the formation of cumulus clouds over or near mountains. Their speeds reach a maximum around 1400 hours.

At night the pressure gradient is reversed. The highlands cool off rather rapidly because of terrestrial radiation losses. Cold and dense air then drains downslope into depressions and valleys. Such cold winds are known as mountain or katabatic (downslope) winds. Such winds are usually cited as the cause of

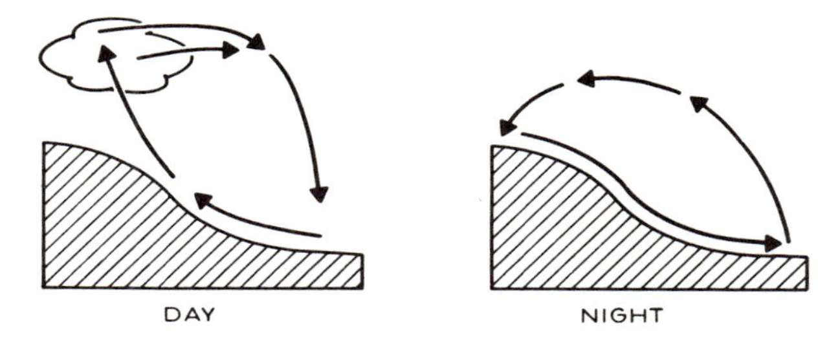

Fig. 5.12. Mountain and valley winds

frost incidence in valleys and depressions of hilly and mountainous areas. They also contribute to the development of temperature inversions in the valley bottoms, a condition that favours the concentration of pollutants in the atmosphere over industrial areas. In temperate regions, the radiational cooling of hill slopes is intensified if the slopes are snow covered. This encourages mass gravity flow of cold dense air into the valley bottoms. This intensifies the incidence of frost and temperature inversion conditions.

Mountain ranges also have effects on winds moving across them. Lee waves or standing waves are formed in the air flow in the lee of mountain barriers if the air is stable. This is because a stable air returns to its original level in the lee of a barrier after being displaced upwards over the obstacle. The descent usually forms the first of a series of waves downwind. Airmen are naturally interested in such phenomena since lee waves and the associated circular air motions known as *rotors* affect aviation. The development of lee waves is usually indicated by the presence of lenticular clouds.

There are other wind systems, mostly local ones and therefore of local importance, which are caused by topography with or without variations in solar heating. These include the *Fohn* or *chinook* and the *bora*. Fohn or chinook is a strong gusty dry and warm wind which develops on the lee side of a mountain range when stable air has been forced to flow over the mountain barrier. Such winds are common in winter and spring in the northern flanks of the Alps and mountains of Central Asia where they are known as the Fohn and on the eastern sides of the Rocky Mountains in North America where they are known as the chinook. In some parts of the world like the northern Adriatic, northern Scandinavia, the northern Black Sea coast and in Japan the winds descending the lee slopes of mountain are cold, despite adiabatic warming. Such winds are known as the bora and they occur mainly in winter when cold continental air masses are forced to rise over a mountain range.

It is quite clear from the above survey that atmospheric motions occur on various time and spatial scales. All these motions also interact and affect one another. Atmospheric motions on all scales—temporal and spatial—affect the weather and climate in any given location although world-wide patterns of weather and climate are primarily controlled by the general circulation of the atmosphere. The weather systems discussed in the next chapter are really component parts of the circulation of the atmosphere. They are circulation systems characterized by particular types of weather.

References

Atkinson, B. W. (1972). The atmosphere. In Bowen, D. Q. (ed.), *A Concise Physical Geography*, Hulton Educational Publications, London.

Barry, R. G. (1967). Models in meteorology and climatology. In Chorley, R. J. and Haggett, P. (eds.), *Models in Geography*. Methuen, London.

Barry, R. G. and Chorley, R. J. (1976). *Atmosphere, Weather and Climate*. (3rd edn), Methuen, London.

Chandler, T. J. (1967). *The Air Around Us*. Aldus, London.

Nieuwolt, S. (1977). *Tropical Climatology*. John Wiley, London.

Palmen, E. (1951). The role of atmospheric disturbances in the general circulation. *Quart. Journ. Roy. Met. Soc.* **77**, 337–354.

Riehl, H. (1969). On the role of the tropics in the general circulation of the atmosphere. *Weather* **24**, 288–308.

Riley, D. and Spolton, L. (1974). *World Weather and Climate*. Cambridge University Press, Cambridge.

Rossby, C. G. (1941). The scientific basis of modern meteorology. *U. S. Department of Agriculture Yearbook Climate and Man*. pp. 599–655.

CHAPTER 6

Weather-producing Systems

Introduction

Weather producing systems are circulation systems accompanied by characteristic weather patterns and types. They cause the day-to-day and week-to-week variations in weather and are often referred to as atmospheric or weather disturbances. These disturbances are extensive waves, eddies, or whirls of air embedded in the general circulation of the atmosphere. The most important of these weather producing systems are the mid-latitude cyclones and anticyclones, the tropical cyclones, and the monsoons. These and other weather disturbances will be discussed in this chapter.

Weather and climate in the middle and high latitudes are largely determined by series of travelling cyclones and anticyclones. A cyclone is the term used to describe atmospheric pressure distribution in which there is a low central pressure relative to the surroundings. Where there is a high central pressure relative to the surroundings the term anticyclone is used. The circulation about the centre of a cyclone is anticlockwise in the northern hemisphere and clockwise in the southern hemisphere and the weather is generally 'stormy.' On the other hand, the circulation about the centre of an anticyclone is clockwise in the northern hemisphere and anticlockwise in the southern hemisphere while the associated weather is generally settled and quiet.

Moving cyclones are of three types:

1 extratropical cyclones typical of middle and high latitudes;
2 tropical cyclones found in low latitudes over ocean areas and adjacent lands; and
3 tornadoes which when they occur over the sea are called water spouts and are referred to as the dust devils in hot arid regions.

The extratropical cyclone is usually called a depression and this term is often preferred to avoid confusion with the tropical cyclone. Most mid-latitude depressions are frontal in origin. Those depressions which are non-frontal are less common and they include polar air depression, thermal depression, and lee depression. But we will consider first the frontal depression which can be described as the motivator of weather in the middle and high latitudes.

Air masses and fronts

Frontal depressions develop only where air masses of contrasting properties exist to encourage frontogenesis—the formation or intensification of fronts. Fronts are boundary zones separating air masses of different properties. The idea of air mass was first introduced into meteorology by T. Bergeron in 1928 while frontal analysis was first introduced by J. and V. Bjerknes in the early 1920s.

An air mass can be defined following Hare (1963) as a large body of horizontal uniform air travelling as a recognizable entity and coming from either a tropical or a polar origin. A simple classification of air masses is given in Table 6.1. The situation is, however, more complex than that presented in the table. This is primarily because the air mass undergoes both thermal and dynamic modifications once it leaves its source region. Thermal modification results from the influence of the thermal characteristics of the surface below the air mass as it moves. Dynamic modification arises from the relations of the air mass with nearby anticyclones and depressions.

Air masses originate in areas where conditions promote the development of vast bodies of uniform and horizontal air. Such areas are usually extensive

Table 6.1 Basic classification of air masses

Major group	Subgroup	Source region	Properties at source
Polar (P) (including Arctic A)	Maritime polar (mP)	Oceans poleward of 50° latitude in both hemispheres	Cool, moist, and unstable
	Continental polar (cP)	1 Continents around the Arctic Circle 2 Antarctica	Cold, dry, and very stable
Tropical (T) including equatorial E	Maritime tropical (MT)	Oceans of the tropics and subtropics	Warm and moist; rather stable on the east side of ocean but unstable on the west side
	Continental tropical (cT)	Low latitude deserts particularly the Sahara and Australian deserts	Hot, very dry, and rather stable

and physically homogeneous. Also, in such areas there must be sufficient stagnation of atmospheric circulation to enable the air mass to acquire the moisture and thermal properties of the underlying surface. Areas of uneven terrain or where land and water are juxtaposed are unsuitable. Similarly areas with predominantly convergent air flow cannot serve as source regions for air masses. The major source regions of world air masses are not only homogeneous but are characterized by anticyclonic circulations which favour the development of horizontal temperature uniformity required in an air mass (Trewartha, 1968). Notable air mass source regions include:

1 the snow-covered arctic plains of North America, Europe, and Asia;
2 the subtropical and tropical oceans;
3 the Sahara Desert of Africa; and
4 the continental interiors of Asia, Europe, and North America.

The longer an air mass remains in its source region before moving the more will it be affected by the thermal and humidity characteristics of the source region. The depth to which an air mass is affected by its source region also depends on the degree of thermal and humidity differences between the air and the underlying surface. As the air mass moves away from its source region it gets modified in its thermal and humidity characteristics in several ways. First, it is influenced by the nature of the surface over which it moves. If the surface is colder than the overlying air the air mass will be cooled from below and will tend to be stable with an inversion or isothermal layer developing in the lowest layers of the air mass. On the other hand, if the surface is warmer than the air mass, the lapse rate will be steepened and the air mass will be relatively unstable. There may also be an increase or decrease in humidity as a result of thermal modification. But significant changes in humidity characteristic occur if the underlying surface is warm and moist or dry and warm. The air mass is thus modified by the different amounts of radiation and moisture which it receives. The air mass is also greatly modified by adiabatic cooling and warming taking place within it. Such processes involve not only condensation and the release of latent heat but also the lifting and subsidence of thick air layers within the air mass.

Air masses are very important in the study of weather and climate because they directly influence the weather and climate over the area in which they prevail. The weather characteristics of an air mass depend on its temperature and humidity characteristics and the vertical distribution of these elements (i.e. lapse rates). The temperature lapse rate will determine the stability or otherwise of the air mass while the moisture content indicates the potential ability of the air mass to yield precipitation.

The major air masses of the world are shown in Fig. 6.1. It is apparent that whereas tropical areas are affected by two or three air masses the middle and high latitudes are affected by three or more air masses. In addition the air masses in the extratropical areas have very contrasting properties unlike the tropical air masses which have more or less similar thermal characteristics and

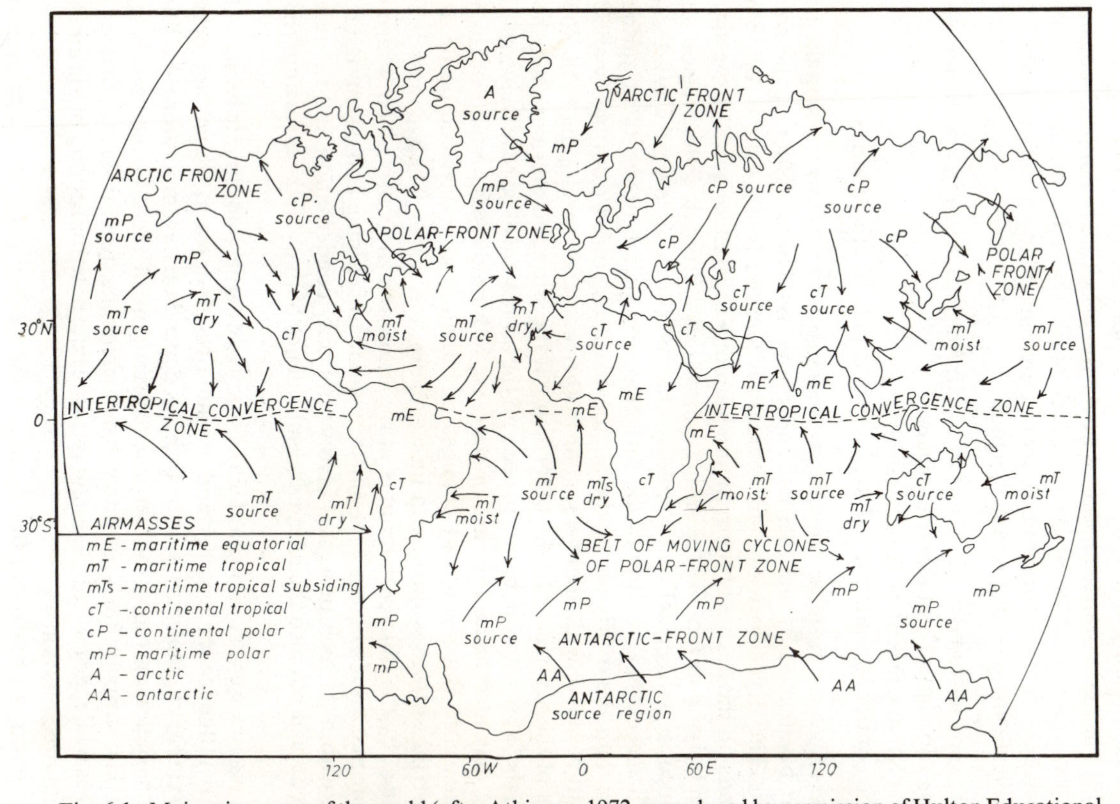

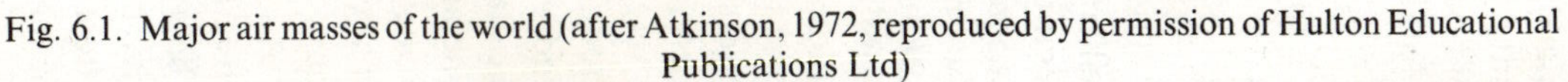

Fig. 6.1. Major air masses of the world (after Atkinson, 1972, reproduced by permission of Hulton Educational Publications Ltd)

differ significantly only in their moisture content. Partly for this reason, weather in the middle and high latitudes is more changeable than weather in the tropics. Also, true fronts do not exist in the tropics and so frontal depressions are absent. The changeability of weather in extratropical areas is primarily due to the effect of travelling sequences of frontal depressions and anticyclones.

Frontal depressions

Three conditions must be fulfilled before frontogenesis (and hence frontal depressions) can take place. First, there must exist two adjacent air masses of contrasting temperatures. Second, there must be an atmospheric circulation with a strong convergent flow to transport the air masses towards each other. Third, there must be sufficient coriolis force to ensure that the warm air does not just lie on the cold one. Whenever the above conditions cease to hold, fronts decay and disappear—a process known as frontolysis. The main areas of frontolysis are Iceland and the Aleutian Islands. The world frontal belt lies roughly between 30th and 60th parallels in both hemispheres. In these zones there are strong poleward temperature gradients throughout the year but these gradients are about twice as strong in winter as in summer. Hence, frontogenesis is more frequent and intense in winter than in summer.

According to the frontal theory of depression formation, depressions usually form as waves on frontal surfaces. Fig. 6.2 shows the six stages in the life cycle of a frontal depression. The first stage is the initial stage in which the front is undisturbed. The second stage marks the beginning of cyclonic circulation with the development of a wave of low amplitude on the front. At the third stage, the warm sector is well defined between the cold and warm sector fronts. The cold front begins to overtake the warm front at the fourth stage and by the fifth stage, there is an occlusion. The warm sector is lifted off and is about to be eliminated. The sixth stage marks the dying stage of the depression. The warm sector has been eliminated and what is left is a whirl of cold air. The life span of a depression is about 4–7 days.

On the synoptic chart, depressions appear as low pressure cells with elliptical isobars. Depressions which are well developed are about 1950 km on the longer axis, 1050 km on the shorter axis. Depressions move eastward from the west at the rate of about 50 km per hour in winter and 30 km per hour in summer. A cross-section of a mature depression is given in Fig. 6.3. There are two types of fronts—the warm front and the cold front. The warm front is the zone where there is an active upglide of the lighter warm air over the denser cold air. The cold front is the zone where there is a forced ascent of warm air over the cold air as a result of the cold air undercutting and lifting the warm air. Fronts range in width from 80 to 240 km. Changes in the weather elements are much more rapid across fronts than within the air masses themselves. The warm and cold fronts differ in several ways. Along the warm front, warm air mass replaces a cooler one whereas the cold front brings about the arrival of colder air. The slope of the cold front is much steeper than that of the warm front, the slope

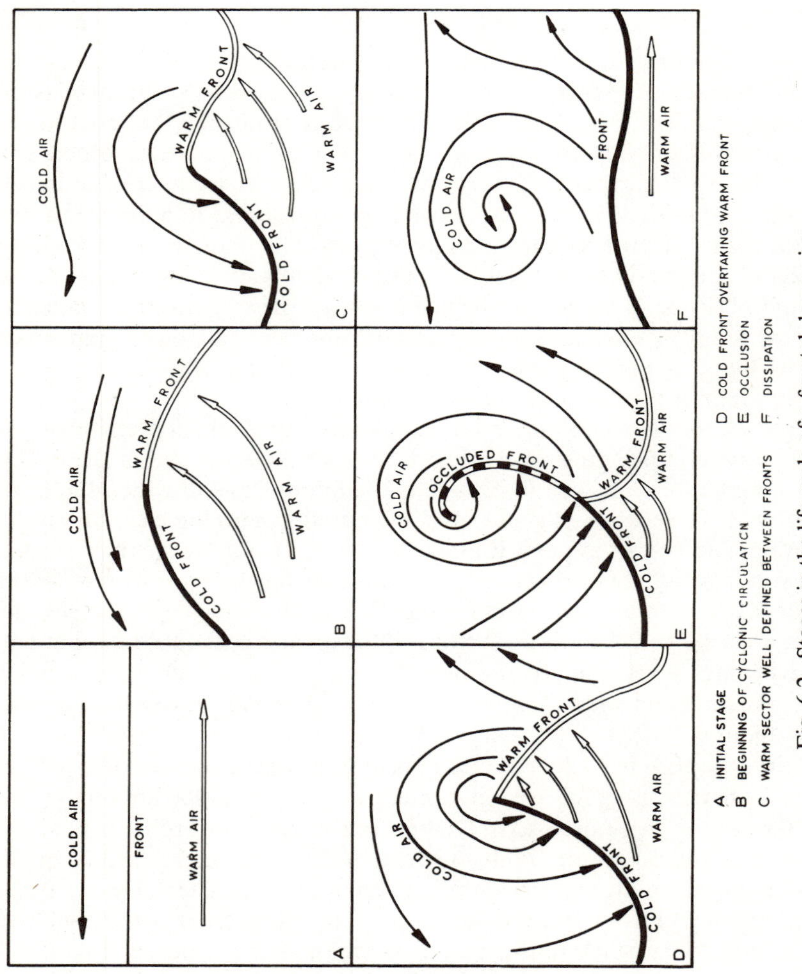

Fig. 6.2. Stages in the life cycle of a frontal depression

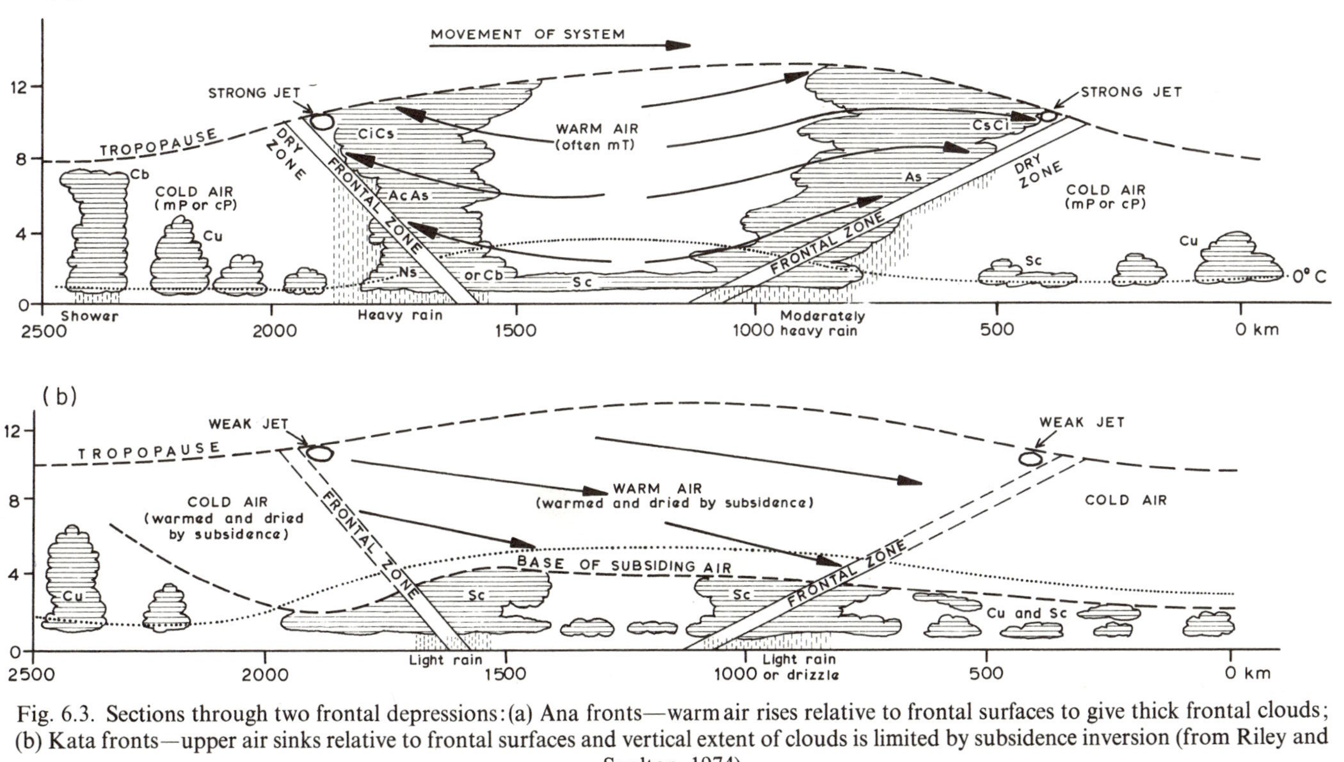

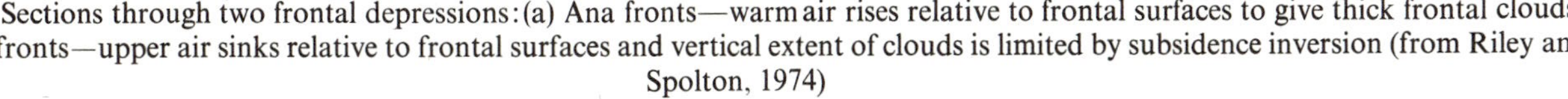

Fig. 6.3. Sections through two frontal depressions: (a) Ana fronts—warm air rises relative to frontal surfaces to give thick frontal clouds; (b) Kata fronts—upper air sinks relative to frontal surfaces and vertical extent of clouds is limited by subsidence inversion (from Riley and Spolton, 1974)

varies from 1 in 40 to 1 in 80 while for the warm front the slope varies from about 1 in 80 to 1 in 200. Fronts move at the rate of about 50–80 km per hour. The cold front is, however, faster than the warm front, a fact that accounts for the occlusion of the warm sector in the later stage of a depression.

Depressions do not generally occur as separate units but in families of three or four. There is the original and primary depression which is succeeded by secondary depressions that form along the trailing edge of an extended cold front. Each new depression follows a course which is south of its progenitor as the polar air pushes farther south to the rear of each depression in the family. Eventually the front trails far to the south and the cold polar air forms a meridional wedge of high pressure that terminates the sequence (Barry and Chorley, 1976). Secondary depressions may also form on the warm front at the point of occlusion where a separate wave forms, running ahead of the parent depression. This development is more likely with very cold air ahead of the warm front and where the eastward movement of occlusion is hampered by mountains.

The normal eastward movement of a depression may also be hampered by the development of stable and stationary anticyclones known as blocking anticyclones. Depressions are forced to move around such blocking highs. A major area of blocking in Europe is Scandinavia, particularly in spring. Another less common area of blocking high in Europe is Iceland. The location of blocking high is of meteorological importance since a blocking high can divert the path of depressions or allow them to stagnate over a given area. The occurrence of blocking patterns is associated with a dominant meridional flow in which pressure difference along the meridian is small and the zonal index is low (see Chapter 5).

The passage of a depression is characterized by the sequence of weather as shown in Table 6.2. Ahead of the warm front, atmospheric pressure falls and the wind backs and increases in speed. Temperature remains steady or rises slowly and there is precipitation. At the warm front, the wind veers and decreases in speed. Temperature rises slowly and there is little or no precipitation. In the rear of the warm front, there is intermittent rain or drizzle. The wind is now steady and there is little or no change in temperature. Ahead of the cold front, pressure falls and the wind backs and increases. There is some rain. At the cold front pressure rises suddenly while the wind too veers suddenly. There is a fall in both temperature and humidity. There is heavy precipitation accompanied by hail and thunder. Behind the cold front, fine spells succeed the rains although these may be followed by further showers. The wind backs slowly and then becomes steady. Humidity is low and there is a slow continuous rise in pressure.

These weather characteristics are obviously not all experienced at a given location when a depression passes overhead for two reasons. First, the characteristics described above and shown in Table 6.2 refer to a mature depression and depressions do age with time. Second, depressions are dynamic systems constantly moving. This means then that a given depression will not stay long enough over a given locality for all these characteristics to be observed. Besides,

Table 6.2 Sequence of weather types accompanying the passage of a depression

Element	In advance	At the front	In the rear
Warm front			
Pressure	Steady fall	Fall ceases	Little change
Wind	Backs and increases	Veers and decreases	Steady
Temperature	Steady or slow rise	Rises slowly	Little change
Humidity	Gradual increase	Rapid rise	Little change
Cloud	Ci, Cs, As, Ns in succession	Low nimbo stratus	Stratus and stratocumulus
Weather	Continuous rain (or snow in winter)	Precipitation almost stops	Fair conditions, or intermittent slight rain or drizzle
Visibility	Good, except in rain	Poor—mist and low cloud produce poor visibility	Often poor, with low cloud and mist or fog
Cold Front			
Pressure	Falls	Sudden rise	Slow, continuous rise
Wind	Backs and increases	Veers suddenly, often accompanied by line squalls, i.e., severe winds along frontal boundary	Slow backing after squall, then steady
Temperature	Steady—sometimes slight fall in rain	Sudden fall	Little change
Humidity	No great change	Sudden fall	Generally low
Cloud	Altocumulus and altostratus, follow-ed by cumulonimbus cumulonimbus	Cumulonimbus, with fracto-culumus or low nimbostratus	Lifts rapidly, but cumulus or cumulo-nimbus may develop
Weather	Some rain, with possible thunder	Heavy rain, often accom-panied by hail and thunder	Heavy rain for a short period; fine spell, followed by further showers
Visibility	Poor—some fog	Temporary deterioration; rapid improvement	Very good

depressions travel in a sequence referred to as a family of depressions. These depressions are separated by areas of higher pressure known as anticyclones. Over a given area, as many as four depressions in various stages of development may pass within a period of 48 hours or less. This then accounts for the complexity and changeability of weather in extratropical areas, particularly in winter then there are more numerous and vigorous depressions compared to the summer season.

Non-frontal depressions

As mentioned earlier, not all depressions are frontal in origin. Some depressions are caused by solar heating while others form in areas of high mountain ranges. The three well known types of non-frontal depressions described here are the thermal, polar air, and lee depressions.

Thermal depressions

These form as a result of prolonged and intense solar heating of the land. Heating causes a general expansion of the air and an outward flow at high levels leading to fall in pressure at ground level. Thermal depressions vary in scale. The Asiatic monsoons can for instance be regarded as large-scale thermal depressions. The thermal depressions are deepest in the afternoon and may disappear altogether at night. Thermal depressions do not cause widespread bad weather unless the air is very damp. In hot deserts, thermal depressions result in hot dry convection winds. In the middle latitudes they are often associated with thunderstorms particularly in summer. Thermal depressions do not usually move and do not persist if cooler weather supervenes. Occasionally they develop into travelling disturbances when they absorb pre-existing fronts.

Polar air depressions

These are depressions which develop wholly within unstable, maritime polar (mP) or Arctic (mA) air. They tend to form south of the centre of an old or occluded frontal depression. Polar air depressions occur mainly in winter. They are small synoptic scale systems with a lifetime of 1–2 days and are highly erratic in motion. Polar air depressions usually brings showery, unsettled weather without marked rain areas. Some of Britain's worst gales have been associated with polar air depressions (Hare, 1963).

Lee depressions

These depressions are associated with high mountain ranges like the Alps, the Rockies, and the Appalachians. When, for instance, a westerly air mass is forced to ascend over a north–south mountain barrier, wave troughs may develop on the lee side of such mountains because of the tendency for convergence and cyclonic curvature there. From such troughs a closed low pressure system may develop depending on the air mass characteristics and the size of the mountain barrier. More than half of the depressions affecting the Mediterranean area of southern Europe and North Africa are lee depressions. Such depressions occur most frequently in winter to the south of the Alps and the Atlas Mountains when this area is under the influence of cold northwesterly air stream. Although fronts may occur in these depressions, the depressions have not formed as waves along frontal zones as in the case of frontal depressions described earlier.

Anticyclones

There are two types of anticyclones: the relatively stationary anticyclones and the travelling anticyclones. The stationary anticyclones are also known as warm anticyclones because they possess a warm core. Temperature throughout the troposphere is abnormally high and the warmth of such anticyclones is maintained through dynamic subsidence. A warm anticyclone intensifies with height. On the other hand, a cold anticyclone weakens with increasing elevation and is replaced aloft by low pressure. The travelling anticyclones are cold anticyclones and are characterized by abnormally cold air in the lower troposphere. These anticyclones are quick-moving, short-lived and shallow, unlike the warm anticyclones which are stable and slow-moving.

Cold anticyclones form mainly in high latitude zones within continental polar or Arctic air. On the other hand, warm anticyclones occur mainly in the subtropical belt especially over the oceans. We are here concerned with the travelling anticyclones since it is these that can be regarded as constituting weather systems.

A travelling anticyclone is normally a high pressure cell, the dynamic counterpart of a depression. Unlike a depression, an anticyclone is usually associated with quiet fair weather which is sometimes spoiled, particularly in winter, by fog or low stratocumulus clouds. Such cloudy overcast conditions are sometimes referred to as the anticyclonic gloom. Some precipitation, mainly drizzle, may fall from such clouds. In summer, however, the weather in an anticyclone tends to be dry, sunny, and warm.

Anticyclones are large weather systems, usually larger than depressions and are characterized by a central region of light winds and of subsidence. Pronounced upward movement required for rain formation is largely absent. Hence they do not give stormy weather like the depressions.

Other systems

Apart from depressions and anticyclones there are other pressure patterns or systems which are of meteorological significance, particularly in the middle and high latitudes. These include ridges, troughs, and cols. A ridge is a region of high pressure in which isobars form the shape of an inverted V between two depressions. A ridge is akin to a spur on a relief map. The weather associated with a ridge is very similar to that associated with an anticyclone. Ridges bring fine or fair weather though this is of much shorter duration than that experienced in an anticyclone.

A trough is the opposite of a ridge and is comparable to a valley on a topographical map. The isobars are V-shaped. A trough is a protuberance from a centre of low pressure and is therefore usually associated with showery weather.

A col is a region of low pressure gradient located between two depressions and two anticyclones. Winds are usually very light within a col. The weather associated with a col is variable, being dependent on the past history of the air in the col and the character of the adjacent pressure systems.

Tropical weather systems

Until the 1940s the only recognized tropical weather system was the tropical cyclone. Measurements taken during World War II in many parts of the tropics indicate the existence of other weather systems. With the advent of meteorological satellites in the 1960s our knowledge of various categories of tropical weather systems has improved considerably.

Tropical disturbances or weather systems can be classified into five categories according to their space and time scales. (Barry and Chorley, 1976). The smallest weather system is the individual cumulus with a life span of only a few hours. The cumulus clouds are generally parallel to the wind direction rather than being randomly distributed. Such cumulus clouds may aid the development of a larger disturbance if convection is intensified by up- and downdraughts during disturbed weather conditions.

The second category is the mesoscale system particularly associated with land/sea boundaries, heated oceanic islands, or topography. Mesoscale systems are intermediate in size and life span between synoptic disturbances and individual cumulonimbus cells. They include thunderstorm and organized convective systems like the squall lines.

The third category is the cloud cluster, a distinctive feature of the tropics that has been identified from satellite imagery. The cloud cluster is of subsynoptic scale and may persist for 1–3 days.

The fourth category includes synoptic scale wave disturbances (e.g. the easterly waves of the Caribbean islands) and cyclonic vortices (e.g. hurricanes and monsoon depressions). The final and fifth category of tropical disturbances is the planetary wave. These waves are very large with a wavelength from 10,000 to 40,000 km and occur in the equatorial upper troposphere and stratosphere (two types). The waves do not, however, appear to influence weather directly though they may interact with lower tropospheric systems.

In this book only those systems which determine much of the disturbed weather of the tropics are discussed. It must also be emphasized that our knowledge of these various weather systems is still incomplete. Many aspects of their development and role in influencing weather and climate are yet to be determined.

Tropical cyclones

A tropical cyclone is an almost circular storm centre of extremely low pressure into which winds spiral. The diameter of the storm ranges from 160 to 650 km and the velocity of the winds varies from a minimum of about 120 to 200 km per hour. The life span of a tropical cyclone is about a week and the storm travels at the rate of 15–30 km per hour. About 50 cyclones occur in a year in the northern hemisphere. Tropical cyclones are very notorious because they cause widespread damage and constitute a serious hazard to shipping and aviation.

Tropical cyclones never originate over land surfaces. In fact, they decay when they move over land or over cool water surfaces. They form over all tropical

Table 6.3 Where and when tropical cyclones occur

Area	Season	Local Name
Caribbean Islands, Pacific coast of Mexico, Florida and south Atlantic coast of USA, Gulf of Mexico	June–October	Hurricanes
China Sea, Philippine Islands, Southern Japan	July–October	Typhoons
North Indian Ocean—Bay of Bengal and Southern India	April–December	Cyclones
South Indian Ocean–Malagasy	November–April	Cyclones
South Pacific Ocean, coast of Northern Australia	December–April	Willy-willies

oceans except the south Atlantic. As shown in Table 6.3, tropical cyclones are known by different names in various parts of the tropics and subtropics where they occur. But wherever they occur, they tend to occur in summer—the warm season. Although the origin of tropical cyclone is not clear, the following conditions favour its development:

1 a large ocean area with surface temperature in excess of 26.7 °C to ensure that the air above is warm and humid;
2 a coriolis force large enough to cause a vortex circulation of air—for this reason cyclones do not form within 5–8° parallels north and south of the equator;
3 a weak vertical wind shear in the basic current—for this reason cyclones form only in latitudes equatorward of the subtropical jet stream characterized by intense wind shear;
4 an upper level outflow above the surface disturbance.

Many cyclones also develop from some pre-existing weak tropical disturbances.

The isobars in a tropical cyclone are nearly circular and the pressure gradients are extremely steep. The pressure near the centre of the cyclone may be as low as 914 mb. Temperature distribution around the storm centre is more or less similar in all directions. No fronts or contrasting air masses are involved in the formation of the tropical cyclone. The source of energy for the maintenance of the tropical cyclone is the latent heat of condensation. Hence the cyclone decays when it moves over land or over cool water surfaces. Most cyclones originate within latitudes 20° north and south of the equator excluding the equatorial belt where the coriolis force is small. They then move westwards or northwestwards (southwestwards in the southern hemisphere). In the northern hemisphere the cyclones may turn northeastwards at about latitude 30–35°. Tropical cyclones are characterized by violent winds and heavy rains. At sea, there is a sea swell that may endanger shipping. On land, the heavy rains often cause flooding and damage to bridges, settlements, and farms.

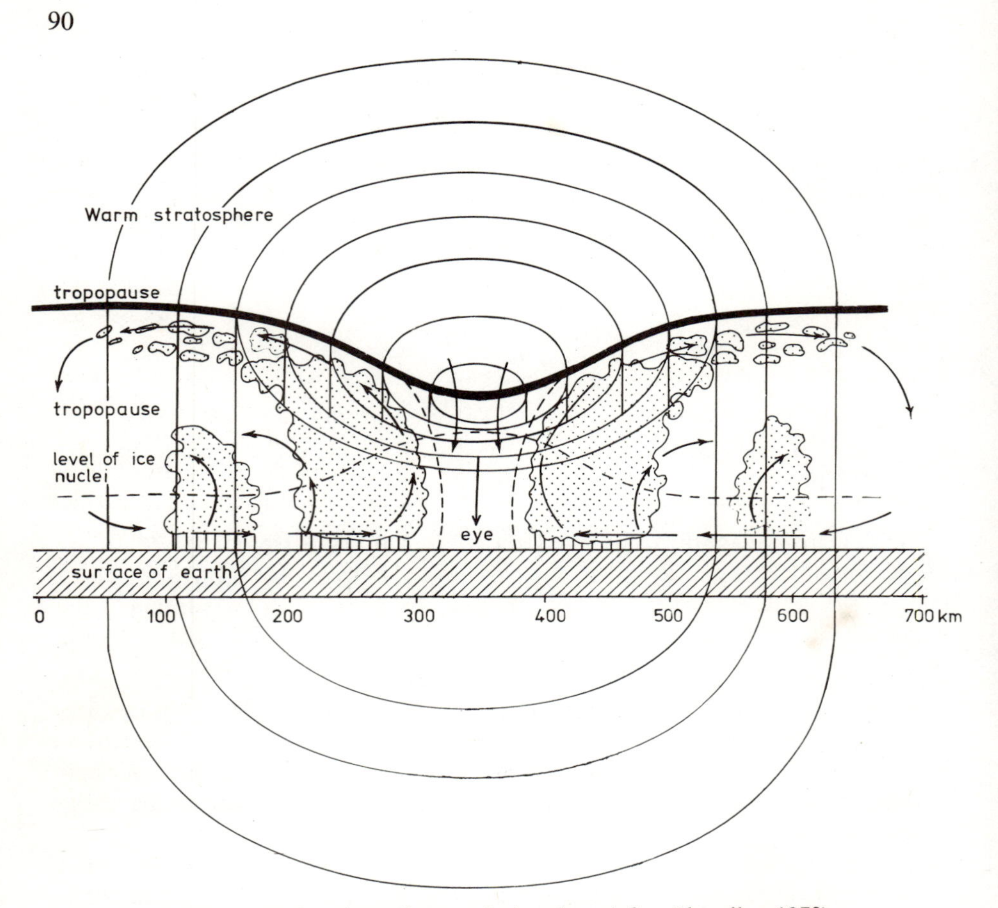

Fig. 6.4. Section through a tropical cyclone (after Chandler, 1972)

In structure, the tropical cyclone consists of two vortices separated by a central calm area known as the 'eye'. For this reason, the stormy weather characterized by violent winds and heavy rainfall is experienced twice. The calm and relatively less cloudy weather of the 'eye' can deceive the unwary into thinking that the storm is over when in fact it is not (see Fig. 6.4).

Tornadoes

A tornado is an extremely intense vortex of small horizontal extent (usually less than 0.5 km) extending downward from a thundercloud. The circulation of wind around a tornado is usually in a counterclockwise (cyclonic) direction. Wind velocities are very high (about 100 metres/s) and are only estimated from the damages caused since no anemometer has survived the passage of a severe tornado. The passage of a tornado is also accompanied by a sudden drop in pressure of about 25 mb which very few buildings can withstand. The intense pressure differential between the outside and the inside of building causes buildings to 'explode' rather than be blown over by the high velocity winds. Tornadoes are easily the most violent of the earth's storms.

The exact origin of tornadoes is not known but they usually occur in conjunction with severe thunderstorms, squall lines, or severe cold fronts. Tornadoes which occur in conjunction with scattered thunderstorms are usually short-lived and have irregular paths. Those which occur in connection with squall lines or severe cold fronts live longer and have more regular and longer paths. Though spectacular and highly destructive along their paths tornadoes are relatively unimportant features of the earth's climate. Tornadoes occur frequently east of the Rockies in the Mississippi Basin in the USA, in eastern India, and east of the Andes. They may occur elsewhere in the world with the exception of the polar regions and the cold northern continents in winter. When tornadoes occur on the sea where they are known as water spouts they constitute a serious hazard to shipping. In hot desert areas, tornadoes only raise dust rather than give any precipitation and they are appropriately called here the dust devils. Tornadoes are therefore not restricted to the tropics and the subtropics although they occur most frequently there.

Weather satellites have since the early 1960s been providing useful information about destructive storms like the tropical cyclones, particularly as regards their occurrences (distribution), structure, and life cycle. Such information, usually in photographic form, has been of help in forecasting the development and track of hurricanes. Weather satellites have been less useful in the case of tornadoes because of their small dimensions. More reliance is put on information from conventional weather stations equipped with storm-detecting radars in addition to other meteorological instruments.

Monsoon depressions

Not all cyclonic systems in the tropics are of the hurricane type. The monsoon depression which affects southern Asia in summer is a less intense type of cyclonic vortex found in the tropics. But it is a very important weather phenomenon as it brings much of the Indian subcontinent 80% or more of its annual precipitation. The monsoon depressions generally move westwards or northwestwards across India, steered by the upper easterlies (Barry and Chorley, 1976). These depressions occur about twice a month when an upper trough is superimposed over a surface disturbance in the Bay of Bengal. Monsoon rains are highly variable from year to year, a testimony to the role played by disturbances in generating rainfall within the moist south westerlies. Breaks also occur within the monsoon rains when the jet stream pushes southwards weakening the Tibetan anticyclone or displacing it northwards. The monsoon depressions of Asia are thus not solely thermal in origin but result from the interaction of both planetary and regional factors at the surface and in the upper troposphere.

Easterly waves

Several types of waves travel westward in the equatorial and tropical tropospheric easterlies with wavelengths of 2000–4000 km and a life span of 1–2 weeks. But the first wave type to be described from the tropics and the best known is the easterly wave of the Caribbean islands (see Fig. 6.5). The wave is characterized by a weak pressure trough which usually slopes eastward with height. Cumulonimbus cloud and thundery showers occur behind the trough line. Easterly waves travel at the rate of 15–20 km per hour. They tend to develop when the trade wind inversion is weak or absent in summer and winter. The penetration of cold fronts into low latitudes may also initiate their development (Barry and Chorley, 1976). They are most frequent during late summer when in the Caribbean area they occur every 3–5 days. This is because at this time the trade wind inversion is weakened by the high surface water temperatures.

The passage of the easterly waves is characterized by the following sequence of weather (Barry and Chorley, 1976). In the ridge ahead of the trough, the weather is fine with some scattered cumulus cloud and a little haze. Close to the trough line, there are occasional showers from the well developed cumulus. These showers help to improve visibility after they have fallen. Behind the trough, the wind veers. There are moderate to heavy thundery showers from

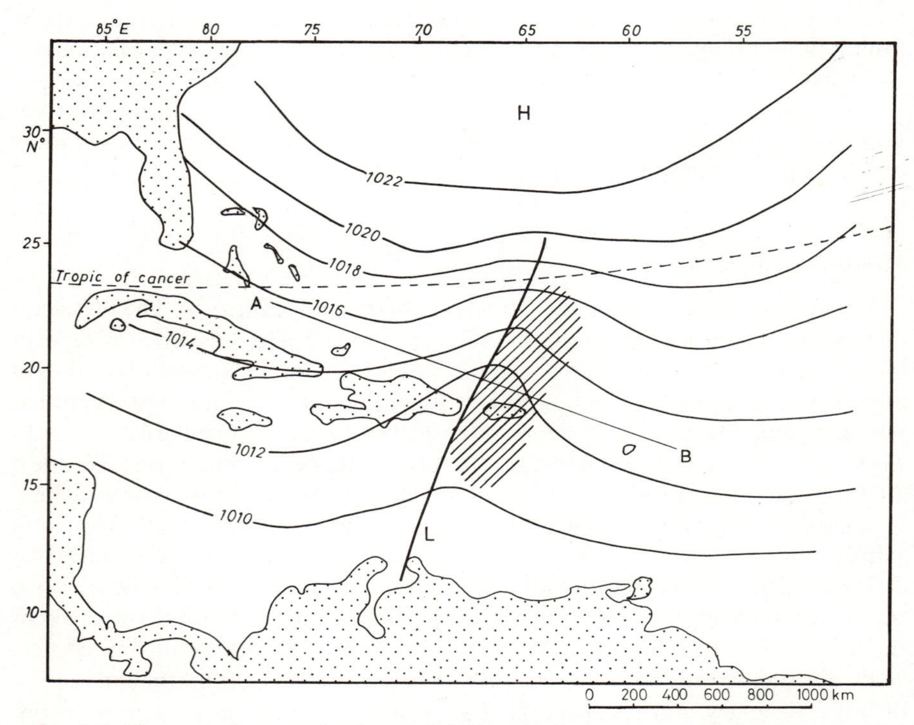

Fig. 6.5. The easterly wave in the Caribbean Islands—hatched area indicates main rainfall zone (from Nieuwolt, 1977)

the heavy cumulus and cumulonimbus clouds. There is a fall in temperature.

The Easterly waves are important for two reasons. First, they bring rainfall to areas which are generally dry as long as the trades remain undisturbed. The late summer rainfall maximum in the Caribbean islands and in the western part of the Pacific is due to the easterly waves.

Second, easterly waves occasionally develop into tropical cyclones and therefore deserve close study.

Linear systems

Linear systems consist of several thunderstorms organized in lines or bands and moving as organized systems. Two well known linear systems in the tropics are the *disturbance lines* of West Africa and the *sumatras* of Malaya. These systems are also sometimes called 'squall lines'. In coastal areas, the sea breeze sometimes develops against the main general wind direction to create a linear system called the 'sea breeze front'. In spite of regional and seasonal variations, the origin of these linear systems is similar being generally due to a combination of various factors of which convergence and convection are the major ones.

The disturbance lines of West Africa have been studied by Hamilton and Archbold (1945) and Eldridge (1957) among others. They are responsible for most of the rainfall in the inland parts of the West African region and the rainfall at the beginning and towards the end of the rainy (monsoon) season in the coastal areas. In other words, the contribution of the summer monsoon rains to the total annual rainfall in West Africa diminishes northward while that of the disturbance lines increases northward. In 1955, for instance, disturbance lines contributed about 30% of the annual rainfall over the coastal areas of Ghana and 90% in the north of the country.

A typical disturbance line is usually 120–240 km long with its axis oriented roughly north–south and moves at the rate of 50 km per hour in a west-southwesterly direction. Ahead of the disturbance line, the surface winds are southwesterlies. As it approaches, a dark cumulonimbus cloud can be observed in the east. As the disturbance line passes, there is a sudden wind shift with strong squally winds from an easterly direction. For an hour or two, easterly winds prevail instead of the southwesterlies. There is a heavy downpour of rain, usually accompanied by strong winds and thunder. The rains are initially of very high intensities and in an hour or two the rains may stop and the surface winds become southwesterly.

Disturbance lines are probably generated by convection over high grounds and they tend to occur in the afternoon. They dissipate over the cold water areas of the North Atlantic. The actual mechanisms underlying the formation of disturbance lines are still not fully understood. Frontal theories such as that of Kendrew (1961) have been denounced. Kendrew (1961) has suggested that disturbance lines are linear frontal disturbances resulting from the interaction of deep monsoon air and the warm moist equatorial easterlies. It is also believed by some people, notably Hare (1963) and Abdul (1966) that the disturbance lines are probably related to the easterly waves since both weather systems are accompanied by similar weather.

The sumatras of southwestern Malaya are another type of linear system caused by solar heating and the effect of orography. They develop during the southwest monsoon season (i.e. from about May to September) normally during the night over the straits of Malacca and move towards the Malayan coast arriving there late in the night or early in the morning. Sumatras consist of a band of cumulus and cumulonimbus clouds measuring 200–300 km in length. Like the disturbance lines of West Africa, they bring heavy rainfall of high intensity but relatively short duration (about 1–2 hours). These rains are accompanied by gusty winds which can reach speeds of up to 70 km per hour.

As in the case of the West African disturbance lines, the origin of the sumatras is not related to any front as they are generally formed within one air stream. According to Nieuwolt (1977) the following factors aid their development.

1 Heating of the southwest monsoon during the day renders the air convectionally unstable and this instability is intensified at night by radiation losses from the top of the cloud when the southwest monsoon is situated over the straits of Malacca.
2 The airmass is uplifted orographically on reaching the Malayan southwest coast.
3 Land breezes from Malaya and Sumatra tend to converge along the straits of Malacca particularly where they are narrow. For this reason, sumatras occur most frequently in the southern parts of the straits of Malacca.

Sumatras are very important because of the rainfall they bring. A single sumatra can for instance give 80 mm of rainfall. The total proportion of rainfall due to the sumatras in southwestern Malaya is therefore considerable (Nieuwolt, 1977).

Thunderstorms

Thunderstorms occur practically everywhere over the globe but they occur most frequently in the tropics (see Fig. 6.6). Also, the intensities of tropical thunderstorms are much higher than those of the middle and upper latitudes. They are therefore of considerable climatological importance in the tropics.

Thunderstorms are highly localized weather phenomena as their diameters are usually less than 25 km. Their duration normally varies from 1 to 2 hours. Thunderstorms develop where there are warm and humid air masses unstable over considerable vertical layers of about 8000 metres. Most thunderstorms are convectional in origin, resulting from intense solar heating but some are caused by sea or land breezes. Orographic lifting along mountain ranges may cause thunderstorms to be distributed in bands or lines called squall lines which may again become organized into linear systems like the disturbance lines earlier described. Thunderstorm showers are sporadic, of short duration, but of very high intensities. The showers are accompanied by squally winds and of course lightning and thunder.

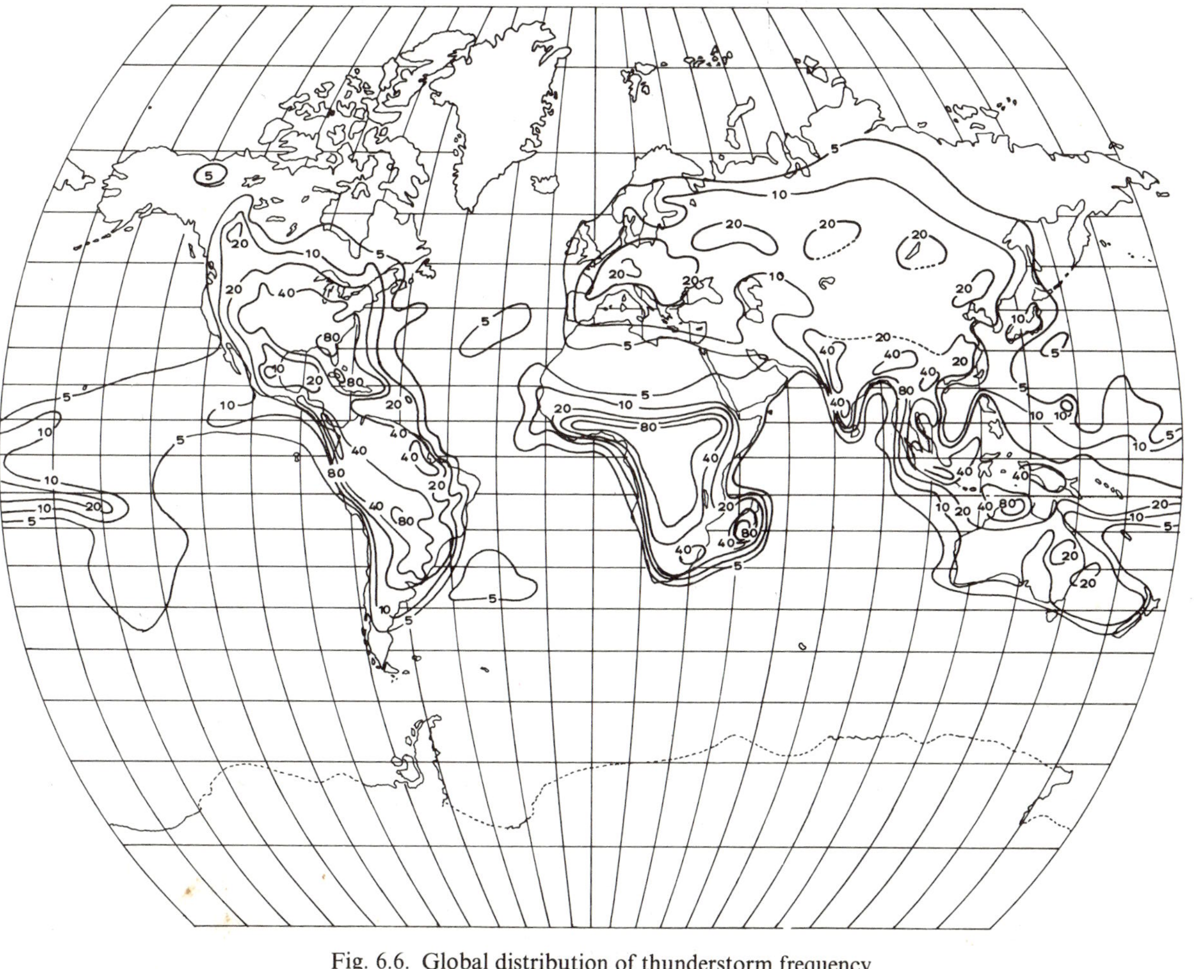

Fig. 6.6. Global distribution of thunderstorm frequency

Lightning is the flash of light accompanying a discharge of atmospheric electricity while thunder is the noise resulting from the sudden heating and expansion of air along the path of the lightning. The origin of lightning is still not completely understood. We know, however, that the earth's surface is negatively charged while the upper atmosphere is positively charged. The normal fair weather electric current is therefore from the atmosphere to the earth's surface. Within a thunder cloud, positive and negative charges tend to collect in different parts as rain drops and ice crystals break into smaller droplets/fragments carrying different charges. When a potential difference of 100 million or more volts is reached there is a spark discharge between the centres of charges. The lightning discharge may be from cloud to ground or from one part of the cloud to another. Along the path of the lightning discharge the air expands so quickly that sound waves are generated. Since light travels faster than sound (300,000 kilometres per second for light and only 330 metres per second for sound) the thunder is heard after the lightning is seen. This time difference can be used to estimate the distance of the storm from an observer. If, for instance, the thunder is heard 5 seconds after the lightning the storm is about 2 km away and if heard 10 seconds after the lightning it is 4 km away and so on in that proportion.

The life cycle of a thunderstorm is very short—about 1–2 hours. Three stages each lasting 20–40 minutes can be recognized (see Fig. 6.7). In the developing stage, strong updraughts prevail in the thunderstorm cell. The cumulus cloud grows rapidly upward to about 8000 metres. There is little or no precipitation and thunder hardly occurs. In the mature stage, the thunderstorm is at its highest intensity. There are some downdraughts even though the updraughts are still strong. There is intense though highly localized precipitation accompanied by thunder. The cumulonimbus cloud may reach up to 18,000 metres and often develops an anvil head caused by upper tropospheric winds (Nieuwolt, 1977). In the dissipating stage, downdraughts prevail. The cloud then exhausts its moisture as the rainfall intensity decreases. Eventually, the cloud dissolves or disintegrates into stratiform clouds.

Thunderstorms occur in distinct patterns thus giving rise to the conventional classification of thunderstorms into three types—air mass thunderstorms, line thunderstorms, and frontal thunderstorms.

1 Air mass thunderstorms are scattered in distribution but occur within the same air mass. They develop locally where the lapse rate has been steepened by intense solar heating. They therefore occur usually in the afternoon and in summer in the middle latitudes.
2 Line thunderstorms are those organized in belts or bands in the direction of winds at the low levels. Such line thunderstorms usually result from the mechanical lifting of a conditionally or convectively unstable air mass by highlands although solar heating may play an additional role in their formation. Such thunderstorms are sometimes called orographic thunderstorms and like air mass thunderstorms they occur in both low and middle latitudes and usually in the afternoon.

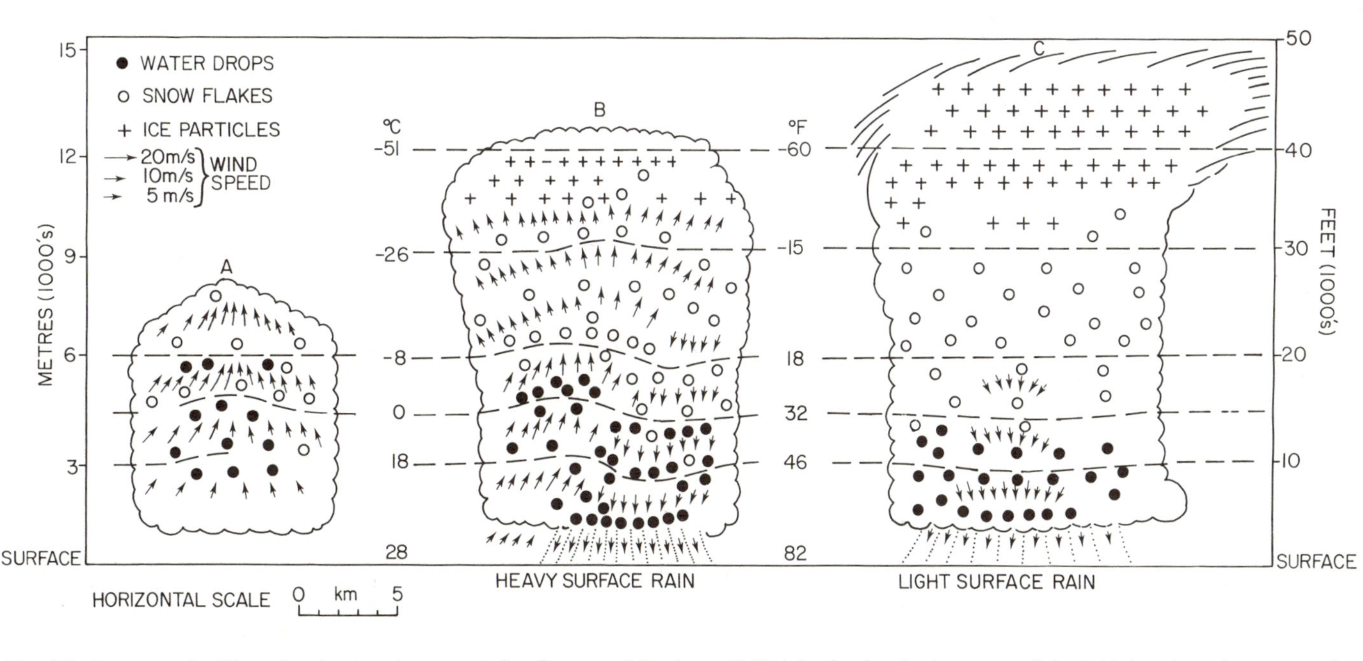

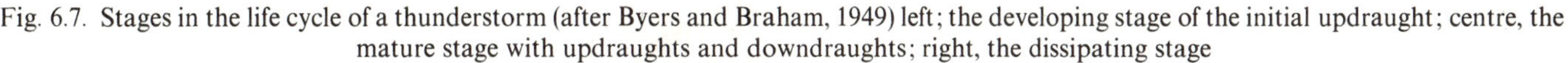
Fig. 6.7. Stages in the life cycle of a thunderstorm (after Byers and Braham, 1949) left; the developing stage of the initial updraught; centre, the mature stage with updraughts and downdraughts; right, the dissipating stage

3 Frontal thunderstorms occur at any time of day or night but only in the middle latitudes along fronts particularly cold fronts. They occur when conditionally or convectively unstable air is forced to ascend over a wedge of colder, denser air. Although frontal thunderstorms may be scattered they move along with the fronts and are organized in their general distribution.

The intertropical convergence zone (ITCZ)

In concluding this chapter, it is necessary to expatiate on the nature and climatological significance of the ITCZ which is sometimes erroneously treated as a tropical weather system. Over most parts of the tropics, the structure of the lower troposphere is characterized by two major air streams: a usually moist but rather cool southerly air with a southwesterly component which forms a wedge under a dry relatively warm air with a northeasterly component. The boundary zone between these two air streams has been given various names by tropical weather analysts. The terms used do reflect the diversity of views on the structure and behaviour of this boundary zone in various parts of the tropics particularly as between land and ocean areas. Such terms as the intertropical front (ITF), intertropical convergence zone (ITCZ), intertropical confluence (ITC), equatorial front, and intertropical discontinuity (ITD) have been used to describe this feature of tropical weather and climate.

It is now established that the boundary zone separating the air masses from the northern and southern hemispheres respectively is neither frontal nor always convergent. The two air streams are too similar, particularly in their thermal properties, for a real front to form between them. Also, convergence between the two air streams is evident usually on the oceans so that on the land the boundary zone between the two air streams is nothing more than a moisture gradient. The term 'intertropical discontinuity' has therefore been proposed by the WMO Provisional Guide to Meteorological Practices to describe the moisture boundary, particularly on land while on the ocean surface the term 'ITCZ' seems appropriate.

The ITCZ/ITD is a rather permanent feature of the tropics whose position is marked by one or more bands of cloud on satellite photographs although sometimes cloudiness may be absent. The structure and characteristics of the ITCZ/ITD vary from region to region depending on such factors as topography and distribution of land and water surfaces among others. It moves northward during the northern summer and moves southward during the southern summer, its mean position being somewhat north of the equator. The range of movement of the ITCZ over the oceans is small but is large over the continent.

In the West African region, the ITD assumes its northernmost position around latitude 20° N in August. This marks the height of the rainy season in West Africa with virtually the whole region under the influence of the moist southwest monsoon from the Atlantic Ocean. The ITD attains its southernmost position around latitude 6° N in January. This is the peak of the dry season in

West Africa with all areas with the exception of the coastal areas under the influence of the dry northeasterlies from the Sahara Desert. There is no particular weather activity associated with the surface position of the ITD over West Africa as such. Its climatological significance lies in the fact that it provides a framework for following the south–north motions of the rain-producing southwest monsoon winds whose depth and motion influence rainfall amount, duration, and distribution. All major weather activities take place on the equatorward side of the ITD in areas located some distance away from the surface position.

In contrast, over the East African region both the summer and winter monsoons are predominantly dry because of their continental origin and the prevalence of low level divergence during both monsoon seasons. Most rainfall in East Africa therefore comes in the transition periods between the monsoons when the low level divergence is temporarily replaced by the more convergent tendencies in the area around the ITD. Consequently, most of East Africa has two rainy seasons in April/May and October/November when the ITD moves over East Africa. Thus, unlike in West Africa or monsoon Asia where either one or both monsoons bring most of the rainfall, rainfall is strongest in East Africa when the monsoons are weakest.

References

Abdul, M. O. (1966). *Summer Climate of West Africa.* Unpublished M. Sc. thesis, University of London.

Alkinson, B. W. (1972). The atmosphere. In Bowen, D. Q. (ed.), *A Concise Physical Geography*. Hulton Educational Publications, London.

Barry, R. G. and Chorley, R. J. (1976). *Atmosphere, Weather and Climate* (3rd edn). Methuen, London.

Byers, H. R. and Braham, R. R. (1949). *The Thunderstorm.* U. S. Government Printing Office.

Chandler, T. J. (1972). *Modern Meteorology and Climatology*. Thomas Nelson, London.

Eldridge, R. H. (1957). A synoptic study of West African disturbance lines. *Quart. J. Royal Met. Soc.* **83**, 303–314

Hamilton, R. A. and Archbold, J. W. (1945). Meteorology of Nigeria and adjacent territory. *Quart. J. Royal Met. Soc.* **71**, 231–262.

Hare, F. K. (1953). *The Restless Atmosphere* (8th impression). Hutchinson, London.

Kendrew, W. G. (1961). *The Climates of the Continents* (5th edn). Clarendon Press, Oxford.

Nieuwolt, S. (1977). *Tropical Climatology*. John Wiley, London.

Riley, D. and Spolton, L. (1974). *World Weather and Climate.* Cambridge University Press, Cambridge.

Trewartha, G. T. (1968). *Introduction to Climate* (4th edn). McGraw-Hill, New York.

CHAPTER 7

Atmospheric Moisture

Significance of atmospheric moisture

Although water vapour represents only 2% of the total mass of the atmosphere and 4% of its volume, it is the most important atmospheric constituent in deciding weather and climate. As pointed out in Chapter 2, the water vapour content of the atmosphere varies widely from place to place and over time at a given location. It can vary from nearly zero in hot arid areas to a maximum of 3% in the middle latitudes and 4% in the humid tropics.

Water vapour is of great significance for several reasons and so meteorologists and climatologists are interested in its amount and distribution over time and space. First, water vapour is the source of all forms of condensation and precipitation. The amount of water vapour in a given volume of air is an indication of the atmosphere's potential capacity to give precipitation. Second, water vapour can absorb both solar and terrestrial radiation and so plays the role of heat regulator within the earth–atmosphere system. In particular it has a great effect on air temperature. Third, latent heat is contained in the water vapour and this energy is released when the vapour condenses. The latent heat contained in the water vapour is an important source of energy for atmospheric circulation and the development of atmospheric disturbances. Fourth, because water vapour contains latent heat its quantity and vertical distribution in the atmosphere indirectly affects the stability or otherwise of the air. Fifth, the amount of water vapour in the air is an important factor influencing the rate of evaporation and evapotranspiration. It is therefore an important determinant of the temperature sensed by the human skin and hence human comfort. Sixth, unlike the other atmospheric gases, water vapour can be changed to liquid or solid form within the range of normal atmospheric temperatures. Water vapour constantly changes phases within the earth–atmosphere system.

Evaporation and evapotranspiration

The atmosphere acquires moisture from the earth's surface through evaporation of water from bare land and water surfaces and transpiration from plants.

Evaporation is the process by which moisture in its liquid or solid form is converted into gaseous form—the water vapour. A distinction is usually made between evaporation and evapotranspiration. The first term is used to describe water loss from water and bare ground surfaces while the latter term is used to describe water loss from vegetated surfaces where transpiration is of major importance. In other words, evapotranspiration is a combined process of evaporation and transpiration.

The rate of evaporation or evapotranspiration over a given area is determined by two major controls. One is the availability of moisture at the evaporating surface while the other is the ability of the atmosphere to vaporize the water and remove and transport the vapour upward. If moisture is always available at the evaporating surface (i.e. non-limiting), then evaporation and evapotranspiration will occur at the maximum rate possible for that environment. This has given rise to the concept of *potential evapotranspiration* discussed later in this book. Moisture is, however, hardly always available in sufficient quantities at the evaporating surface so that evaporation and evapotranspiration often occur at rates below those that would take place assuming water was always available. This has given rise to the idea of *actual evapotranspiration.*

The second major control of evaporation and evapotranspiration mentioned above is a function of several factors including solar radiation, temperature, wind speed, and humidity. Energy is required to vaporize water (about 590 calories for one gram) and the availability of energy is indicated by air temperature in the absence of radiation data. Air turbulence (wind speed) ensures that the moistened air lying over the evaporating surface is removed and replaced by fresh relatively dry air to maintain the evaporation process. The humidity of the air exerts an influence on the rate of evaporation because it is this factor that determines the capacity of the air to hold moisture. The lower the humidity the greater is the capacity of air to hold moisture. Also, there must be a vapour pressure gradient between the evaporating surface and the overlying air since evaporation can only take place if the vapour pressure at the surface is greater than that of the air above it. Consequently, low humidity encourages evaporation while high humidity suppresses it.

Available data on measured evaporation are scanty and often unreliable. Evaporation is measured with the aid of an evaporation tank of which there are different sizes and shapes. The US Weather Bureau Class A Pan has been adopted by the WMO as the standard pan for evaporation measurement. This is a white cylindrical pan about 1200 mm in diameter and 250 mm deep. It is filled with water to within 50 mm of the rim and mounted on a wooden platform on the ground so that the water surface is about 300 mm above the ground. Because of its relatively small surface area, the absorption of radiation by the walls of the pan and the fact that the pan creates its own aerodynamic environment being exposed above the ground, the evaporation values obtained are slightly exaggerated. This phenomenon is referred to as the 'oasis effect', implying that evaporation pans behave as oases within their environment so that most of the energy available around is used to vaporize water from the pans.

The evaporation from large water surfaces is estimated to be about 70–75% of that from a Class A pan in the same environment. Similarly, the evaporation from a wet bare soil is estimated to be about 90% of that from an open water surface since water is less easily released for evaporation from soil compared to the open water surface.

Rates of evaporation or evapotranspiration can also be estimated using formulae. Many of these formulae are empirical and do not therefore give reasonable estimates in areas outside the locations or in dissimilar locations for which they have been derived. The formulae which are theoretical in nature are applicable anywhere though they have their own limitations. Theoretical evaporation formulae are based on two fundamental approaches—the aerodynamic approach and the energy budget approach. In the aerodynamic approach, the vapour flux from an evaporating surface is related to processes of turbulent diffusion. Evaporation is expressed as a function of wind speed and vapour pressure gradient over the evaporating surface. This requires the arduous and difficult task of measurements of wind velocity and vapour pressure at or near the evaporating surface and at another level above it. A good example of such an aerodynamic evaporation formula is that of Thornthwaite and Holzman (1939) given as

$$E = \frac{17.1(e_1 - e_2)(u_2 - u_1)}{T + 459.4} \tag{7.1}$$

where E is evapotranspiration off a short vegetation, e_1 and e_2 are vapour pressures in inches of mercury (in Hg) at height of 2 ft and 28.6 ft above the ground, u_1 and u_2 are the wind velocities in miles per hour at the same two levels and T is air temperature in °F.

In the energy budget approach, the problem resolves into determining the proportion of the net radiation utilized in vaporizing water. If we ignore the other components of the energy budget equation like heat used for heating the ground and for photosynthesis, the net radiation can be partitioned between sensible heat and latent of evaporation using the following formula by Bowen (1926):

$$\beta = \frac{H}{LE} = 0.659\left(\frac{Kh}{Ke}\right)\left(\frac{Ts - Ta}{es - ed}\right) \tag{7.2}$$

where β = Bowen ratio, H is sensible heat (i.e. heat flux to the air), LE is energy used for evapotranspiration, Kh and Ke are eddy diffusivities for heat and vapour respectively, Ts and es are the temperature and vapour pressure at the evaporating surface and Ta and ed are the air temperature and vapour pressure respectively. If we assume that $Kh = Ke$, β can be solved by measurements of temperature and vapour pressure gradients. The amount of evaporation or evapotranspiration can then be calculated from the equation of the form

$$E = \frac{Rn - H}{1 + \beta} \tag{7.3}$$

where E is evapotranspiration, Rn is the net radiation and β and H are as previously defined.

This formula has some limitations. First, the other components of the energy budget equation, namely heat flux to the ground and heat for photosynthesis, are ignored. Second, the assumption that $Kh = Ke$ is unrealistic particularly under conditions of instability (Chang, 1964). Third, the Bowen's ratio is difficult to determine as this requires continuous measurements of profiles of temperature and vapour pressure (humidity) over the evaporating surface as in the case of aerodynamic evaporation formula.

Penman (1948) has provided a formula which is based on a clever combination of the energy budget and aerodynamic approaches and does not require the rather difficult measurements of temperature and humidity profiles over the evaporating surface. Penman's formula is of the form:

$$Eo = \frac{\Delta Rn + \gamma Ea}{\Delta + \gamma} \tag{7.4}$$

where Eo is open water evaporation, Rn is the net radiation, Δ is the slope of the saturation vapour pressure curve for water at the mean air temperature in mmHg/°F, γ is the constant of the wet and dry bulb psychrometer equation (0.27 mmHg/°F) and Ea is an aerodynamic term. This term is computed using the equation of the form:

$$Ea = 0.35\,(ea - ed)\quad(1 + u_2/100) \tag{7.5}$$

where ea is the saturation vapour pressure of water at the mean air temperature in mmHg, ed is the actual vapour pressure of water at the mean air temperature and u_2 is wind speed in miles per day at two metres above the ground.

Penman's evaporation formula is widely used and is generally recognized as the best available. However, for large water bodies account must be taken of heat storage within the water bodies otherwise the formula will underestimate rates of evaporation. Values of potential evapotranspiration (PE) can also be obtained using the above formula by multiplying computed values of Eo by a seasonal reduction factor (f) whose value varies from 0.6 to 0.8 (see Table 7.1).

The major drawback of Penman's formula is that it demands a lot of data

Table 7.1 Values of seasonal conversion factor (f) for estimating values of potential evapotranspiration (PE) from those of open water evaporation (Eo) (after Penman, 1948)

	f
May–August	0.8
September and October	0.7
November–February	0.6
March and April	0.7
Year	0.75

which may not be readily available particularly in the developing countries of the tropics. Data on net radiation are particularly scarce and these often have to be estimated using empirically derived formulae thus increasing the magnitude of errors in the final evaporation estimates. For these reasons, Thornthwaite's formula is often used instead of the more accurate and theoretically based Penman's formula.

Thornthwaite's formula (Thornthwaite, 1948) is really for estimating values of potential evapotranspiration. The formula is highly empirical, lacking in mathematical elegance and without a nomograph would have been very difficult to apply. The formula assumes that evaporation is nil when the mean air temperature is less than 0 °C—a very questionable assumption. Vital meteorological controls of rates of evaporation and evapotranspiration such as radiation, humidity, and wind speed are not included in the formula. In Thornthwaite's formula, potential evapotranspiration is expressed as an exponential function of the mean monthly air temperature. A day length factor is then applied to correct for season and latitude as shown below

$$PE^* = 1.6(10\,T/I)^m \tag{7.6}$$

where PE^* is unadjusted potential evapotranspiration based on a 12 hour day and 30 day month, T is mean monthly air temperature in °F, I is the annual heat index and m is a cubic function of I empirically determined.* The annual heat index (I) is a summation of the values of the monthly heat index (i). Thus

$$I = \sum_{i=1}^{12} i \quad \text{and} \quad i = \left(\frac{T}{5}\right)^{1.514} \tag{7.7}$$

where T is the mean monthly temperature in °F. Finally, an adjustment is made for the actual day length (h) and days in the month (D), since the day length varies with season and latitude and the month is not always 30 days long. The final formula is therefore as follows:

$$PE = PE^*\left(\frac{h}{12}\right)(D/30) \tag{7.8}$$

Several modifications have been suggested by various workers to improve the accuracy of Thornthwaite's formula for estimating values of potential evapotranspiration (see Garnier, 1956 and Sibbons 1962). These modifications seek to include the humidity factor in the equation. Because of the lack of this factor in the equation, the formula tends to underestimate values of potential evapotranspiration in arid and semi-arid areas and during the dry season in seasonally humid environments.

Distribution of evaporation

Considering the important role of energy supply and availability of water in determining rates of evaporation, it is not surprising that evaporation is general-

*$m = (0.6751^3 - 771^2 + 129{,}201 + 492{,}390) \times 10^{-6}$ (see equation 7.6).

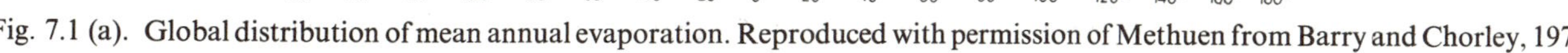

Fig. 7.1 (a). Global distribution of mean annual evaporation. Reproduced with permission of Methuen from Barry and Chorley, 1976

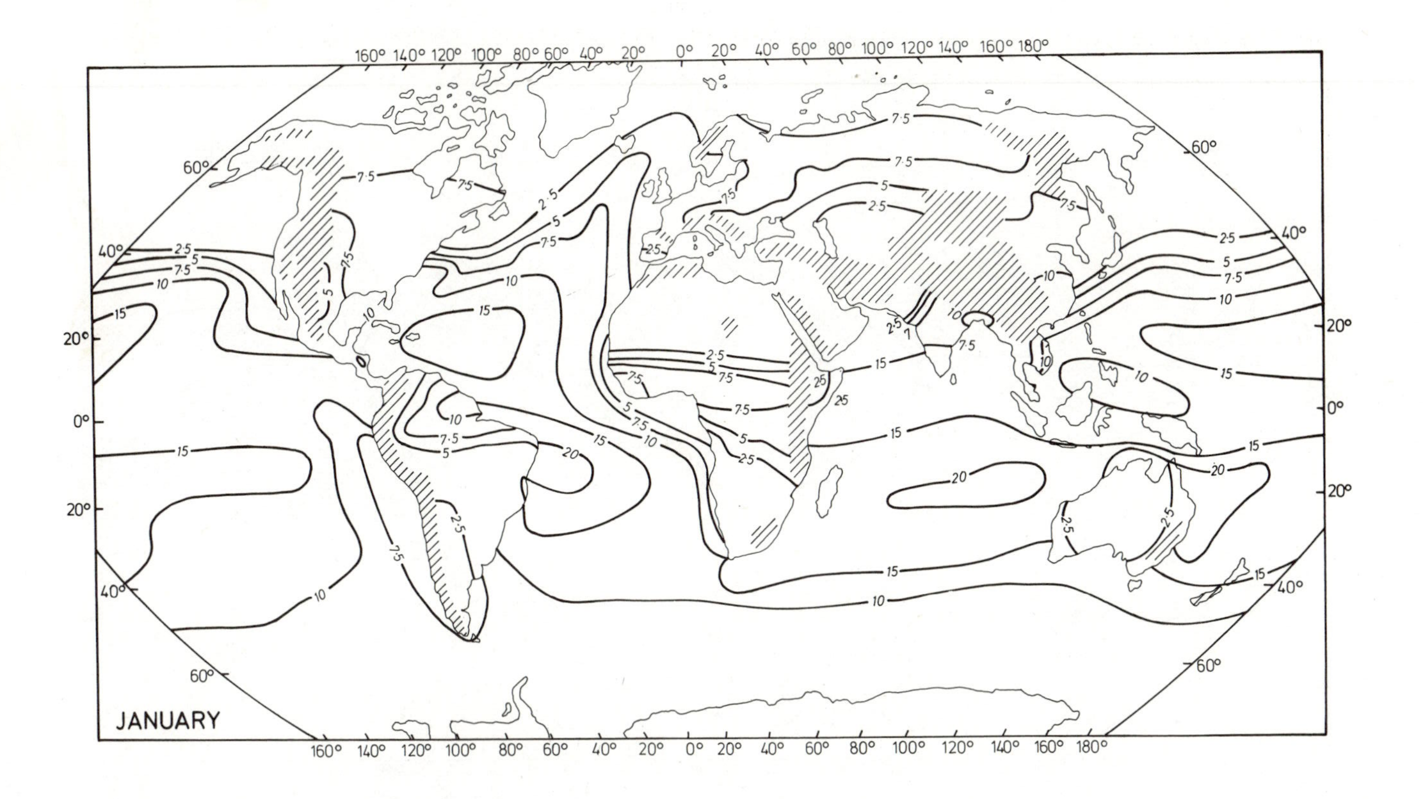
160° 140° 120° 100° 80° 60° 40° 20° 0° 20° 40° 60° 80° 100° 120° 140° 160° 180°
20° 0° 20°
40° 60°
JANUARY

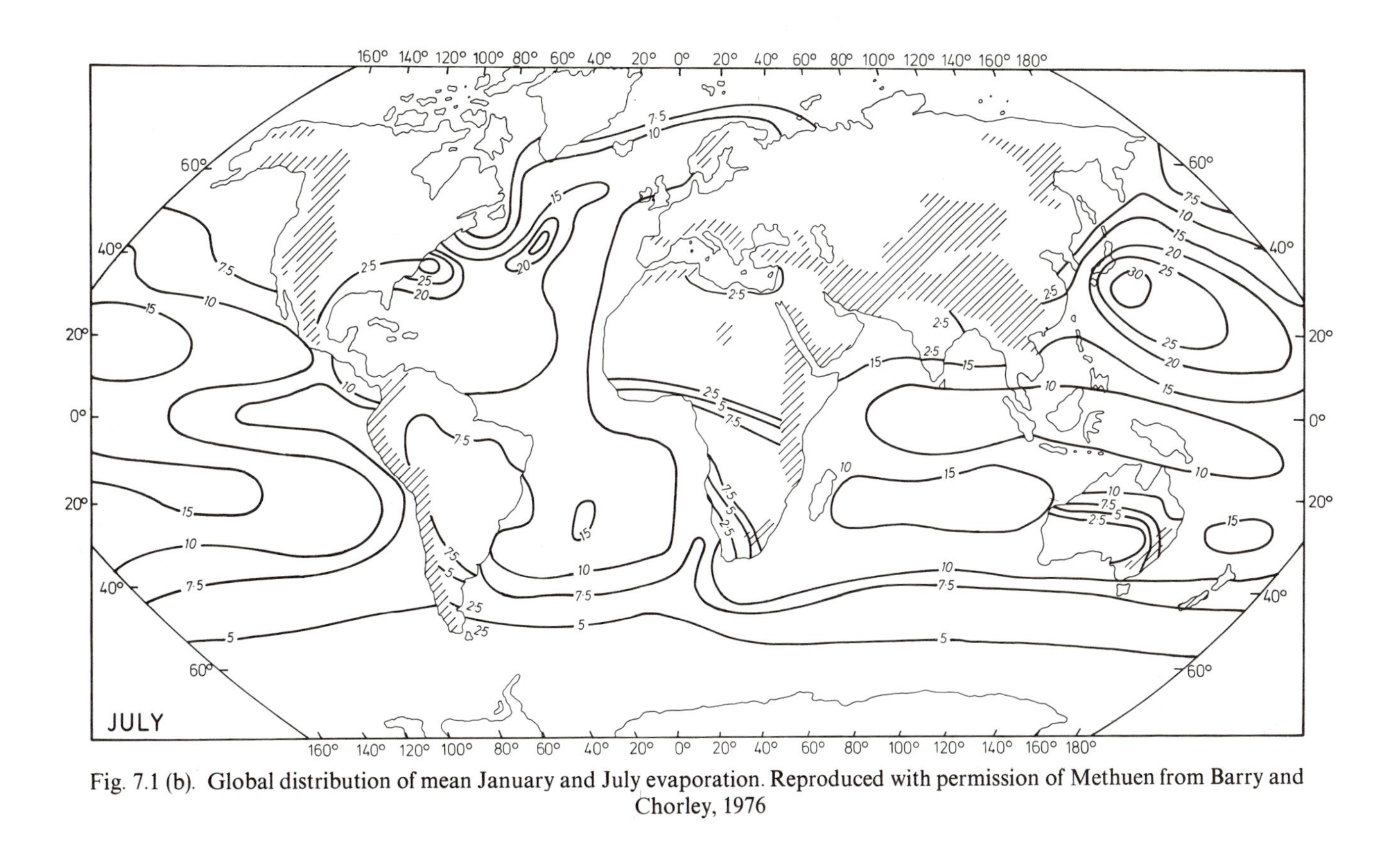

Fig. 7.1 (b). Global distribution of mean January and July evaporation. Reproduced with permission of Methuen from Barry and Chorley, 1976

ly greater over oceans than over land and in the low latitudes than in the middle and high latitudes (see Fig. 7.1). On an annual basis, maximum evaporation losses occur on the oceans located about 15–20 °N and 10–20 °S in the trade wind belt. Evaporation values over oceans in the equatorial zone are slightly lower for three reasons. First, the equatorial zone winds (doldrum belt) have lower velocities than the trades. Second, the equatorial air has a vapour pressure close to the saturation point so that the relative humidity is high. Third, insolation around the equatorial zone is lower than in the trade wind belt because of more frequent and greater cloud cover.

The maximum values of evaporation over land, however, occur around the equator owing to the relatively high insolation receipts and the large transpiration losses from the luxuriant vegetation. The evaporation losses over land in the middle latitudes are also considerable owing to the strong prevailing westerly winds (Barry and Chorley, 1976).

Fig. 7.2 shows the average annual latitudinal distribution of evaporation (E), precipitation (r) and total runoff. It indicates that precipitation exceeds evaporation poleward of 40° and between 10 °N and 10 °S. The resultant runoff from these regions must replenish the excess evaporation of the subtropical

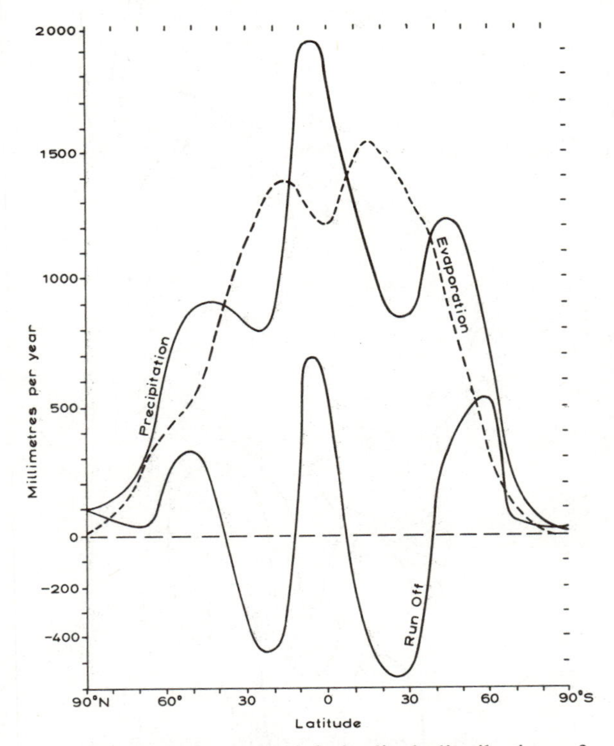

Fig. 7.2. Average annual latitudinal distribution of evaporation, precipitation, and runoff. Reproduced with permission from Sellers, *Physical Climatology*, 1965. Published by The University of Chicago Press

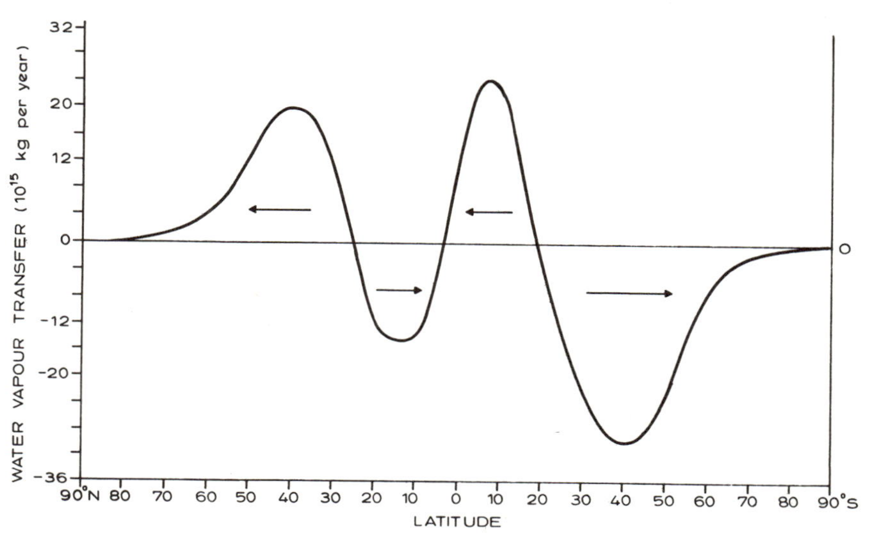

Fig. 7.3. Latitudinal distribution of the average annual meridional transfer of water vapour in the atmosphere (after Sellers, 1965)

belts. Although there are still some uncertainties arising from poor knowledge of precipitation and evaporation rates in the low latitudes three features of the global pattern of vapour transfer can be identified (see Fig. 7.3):

1 The vapour flux is poleward north of 20 °N and south of 20 °S.
2 The magnitude of the vapor flux between 35 and 40 °S is about 50% greater than that between 35 and 45 °N.
3 There is a net flux of water vapour into the intertropical convergence zone (ITCZ) from both the north and south.

Humidity

Humidity is the term used to describe the water vapour content of the atmosphere. It does not cover the other forms in which moisture may be present in the atmosphere, e.g. liquid form (water droplets) and solid form (ice). Atmospheric water vapour originates from the earth's surface by evaporation and transpiration. It is therefore strongly concentrated in the lower layers of the atmosphere..In fact, about half of the total water vapour in the atmosphere is found below 2000 metres. There is usually a steady decrease in the moisture content of the atmosphere with increasing height. Beyond the tropopause, water vapour is virtually absent. Table 7.2 shows the average vertical distribution of water vapour in the middle latitudes. Similar data for the tropics are not readily available because of the sparseness of radiosonde stations. The distribution patterns of the mean atmospheric water vapour content over the earth in January and June are shown in Fig. 7.4. Fig. 7.4(a) shows that in January the vapour content of the atmosphere is highest in the equatorial zone, particularly

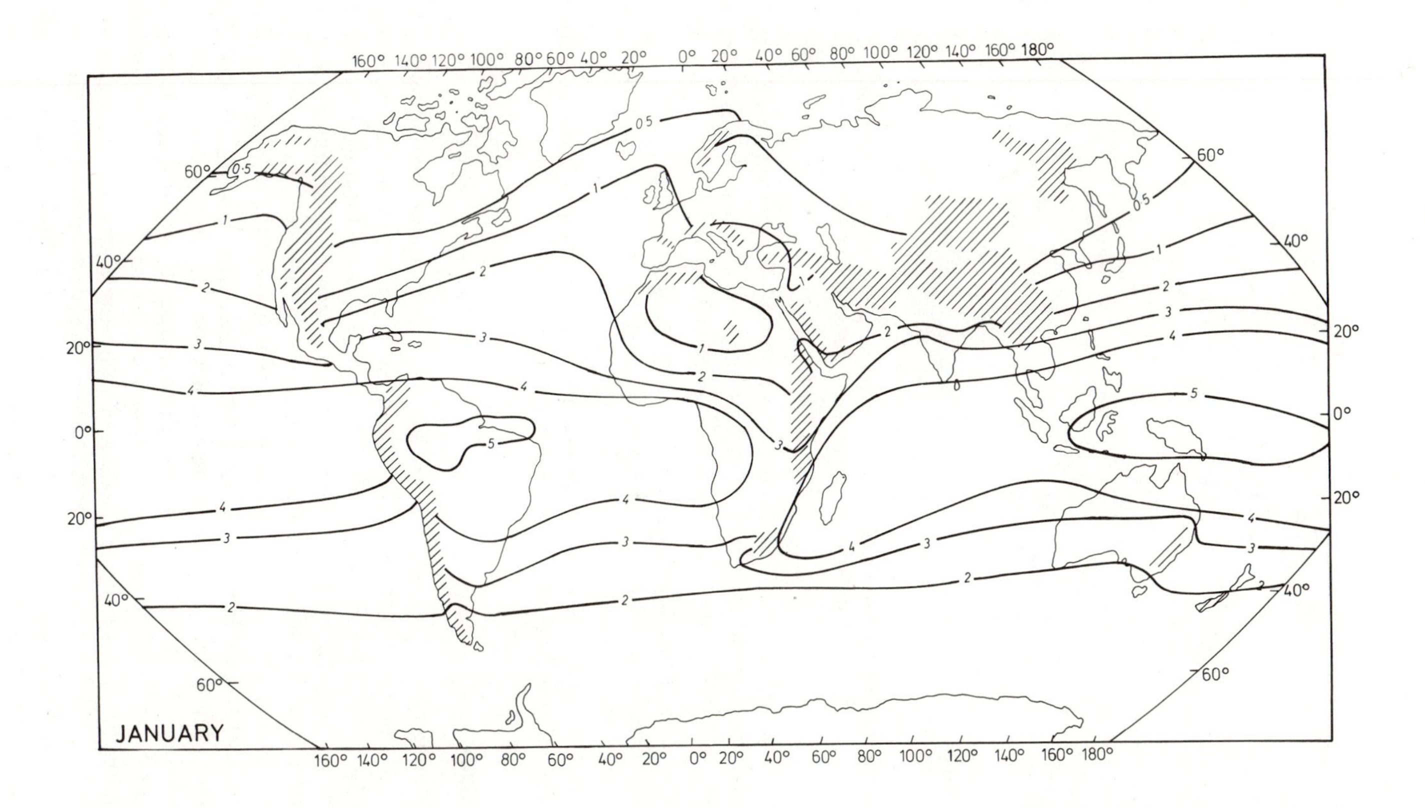
JANUARY

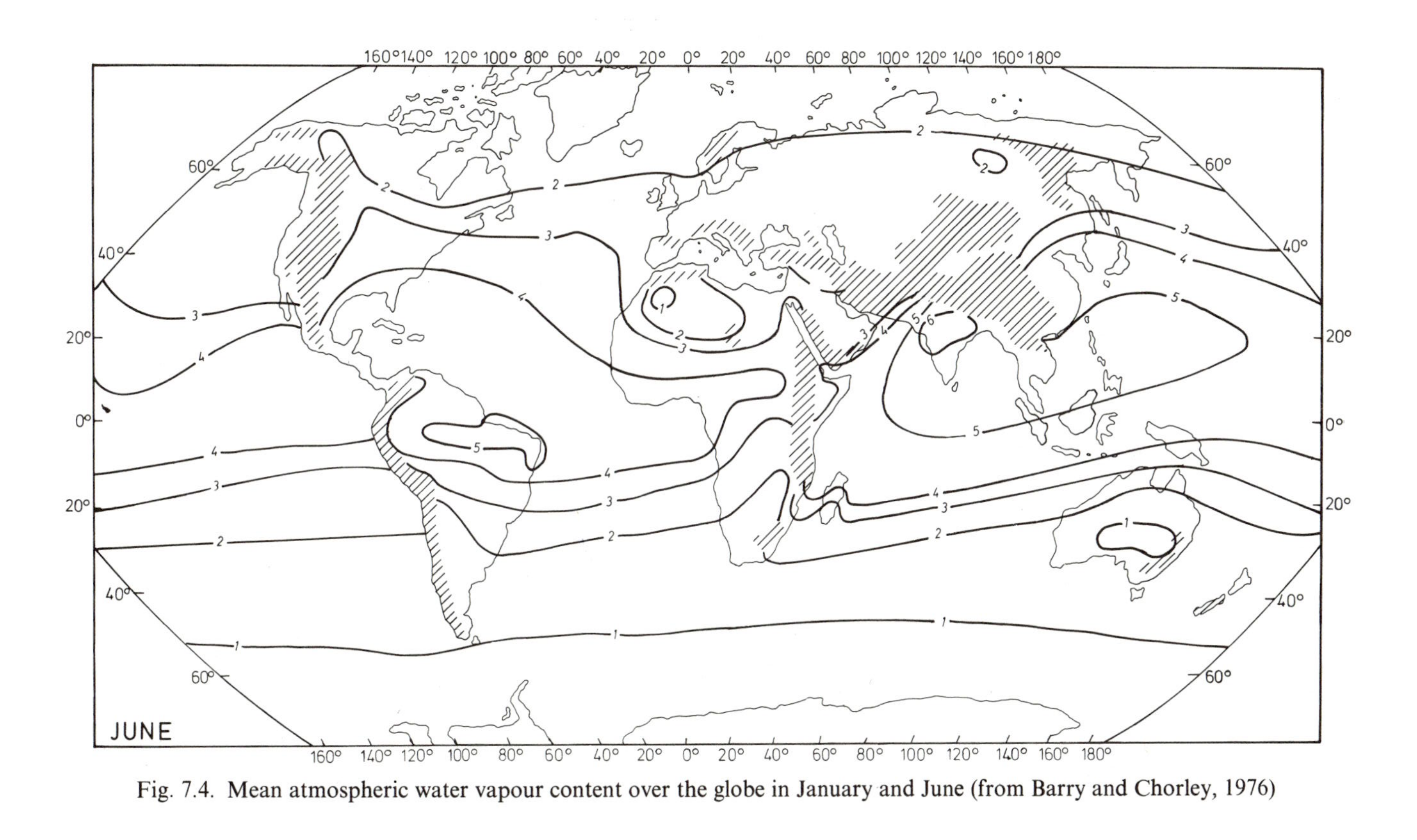

Fig. 7.4. Mean atmospheric water vapour content over the globe in January and June (from Barry and Chorley, 1976)

Table 7.2 Average vertical distribution of water vapour in the middle latitudes (after Landsberg, 1966)

Height (km)	Water vapour (% volume)
0.0	1.30
0.5	1.16
1.0	1.01
1.5	0.81
2.0	0.69
2.5	0.61
3.0	0.49
3.5	0.41
4.0	0.37
5.0	0.27
6.0	0.15
7.0	0.09
8.0	0.05

over the oceans and decreases gently towards the south but more rapidly towards the northern pole. The June map shows a similar distribution pattern except that the middle latitudes in the northern hemisphere have more vapour than similar latitudinal locations in the southern hemisphere which is now in its winter season. The patterns shown in Fig. 7.4 are not surprising in view of the fact that the capacity of an air parcel to hold moisture increases with increasing temperature (see Table 7.3). Consequently, provided water is available for evaporation and the air is warm the water vapour content of the atmosphere will be high. On the other hand, if the underlying surface is dry and/or the air is cool, the water vapour content of the atmosphere will be low. In Fig. 7.4 the highest values of atmospheric vapour of 5–6 cm are to be found over southern Asia during the summer compared to less than 2 cm in the Sahara and other

Table 7.3 Values of moisture content at saturation point for given temperatures (after Gates, 1972)

Temperature (°C)	Moisture content (g/m^3)
− 15	1.6
− 10	2.3
− 5	3.4
0	4.8
10	9.4
15	12.8
20	17.3
25	22.9
30	30.3
35	39.6
40	50.6

deserts during the same season. The lowest values of less than 5 mm are to be found over high latitudes and the continental interiors of the northern hemisphere in winter.

There are various ways of measuring the moisture content of the atmosphere. The following are the indices of humidity usually used.

1 *Absolute humidity* which is expressed in grams per cubic metre of air is the total mass of water in a given volume of air.
2 *Specific humidity* is the mass of water vapour per kilogram of air including its moisture.
3 *Mass mixing ratio* or *humidity mixing ratio* is the mass of water vapour per kilogram of dry air.
4 *Relative humidity* is the ratio of actual moisture content of a sample of air to that the same volume of air can hold at the same temperature and pressure when saturated. It is usually expressed in percentage form.
5 *Dew-point temperature* is the temperature at which saturation will occur if the air is cooled at constant pressure without addition or removal of vapour.
6 *Vapour pressure* is the pressure exerted by the vapour content of the atmosphere in millibars.

The relative humidity is the most popularly used measure of air humidity because it is easily understood and computed using wet and dry bulb thermometers. Besides, it indicates the degree of saturation of the air. The relative humidity is, however, greatly influenced by air temperature. The value can change if there is a change in air temperature even though there has not been any increase or decrease in its moisture content. For instance, the relative humidity of the air varies inversely with temperature, being lower in the early afternoon and higher at night. It is important always to bear in mind that relative humidity does not really give information about the quantity of moisture in the atmosphere but on how close to saturation the air is.

In the tropics where diurnal variations in temperature are large there are considerable variations in humidity in the course of the day. In many parts of the humid tropics, particularly the coastlands, the relative humidity may often be close to 100% at nights during the rainy season. Relative humidity reaches its lowest value in the afternoon during the dry season in the continental interiors of the tropics. There are also seasonal variations in the values of the relative humidity in the low latitudes. Seasonal variations are smallest at the equator and increase with increasing latitude.

Because of the dependence of the values of relative humidity on air temperature, relative humidity data are strictly not comparable for different stations unless they have been obtained at about the same hour of the day and when air temperatures are not too different. For comparison purposes other indicators of atmospheric moisture such as the vapour pressure or the absolute humidity should be used. These measures unlike the relative humidity are not unduly influenced by air temperatures.

Condensation

Condensation is the process by which water vapour is transformed into liquid water. Condensation occurs under varying conditions associated with changes in one or more of air volume, temperature, pressure, or humidity. Condensation thus takes place (see Barry and Chorley, 1976):

1 when the air is cooled to its dew-point even though the volume remains constant;
2 if the volume of the air is increased without addition of heat so that the air is cooled by adiabatic expansion;
3 when a joint change of temperature and volume reduces the moisture-holding capacity of the air below its existing moisture content.

In the atmosphere, condensation usually takes place when the air is cooled beyond its dew-point. As mentioned earlier, the capacity of the air to hold moisture in vapour form decreases with decrease in its temperature. Cooling of the air is the normal method to achieve saturation and hence condensation. Such cooling may be brought in any of the following ways:

1 loss of heat by conduction to a cold surface, a process known as contact cooling;
2 mixing with colder air;
3 adiabatic cooling due to ascent of air.

Contact cooling will give rise to dew, fog, or frost when the water vapour condenses. These condensation phenomena are described later in this chapter. Contact cooling is usually produced within warm moist air when it passes over a cold land surface or even a cold water surface. The frequent incidence of fog off the Ghana and Namibian coasts in Africa is due to warm moist air moving over the relatively cool sea surfaces in these areas. Contact cooling also occurs in the air on clear calm nights when strong terrestrial radiation is favoured. As a result of this, the earth's surface is quickly cooled and this surface cooling will eventually extend to the overlying air. If the air is moist enough and cooling goes beyond the dew-point, condensation occurs in form of dew, fog, or frost depending on the amount of moisture in the air layer, the thickness of the cooling air layer, and the dew-point value (Barry and Chorley, 1976).

Cooling may also result when air masses of different temperatures mix or differing layers within the same air mass mix. Fog or low stratus cloud may result. If the cloud is thick enough some drizzle may fall. The most effective cooling mechanism is, however, the dynamic process of adiabatic cooling following air mass ascent. A given air parcel may rise if it is warm and buoyant (i.e. lighter than the air around). Alternatively, it may be forced to rise in one or both of two ways. It may be forced to rise by orographic barrier or by the effect of colder air pushing the warm one upwards at fronts.

Cooling of air beyond its dew-point is not sufficient to bring about condensation in the atmosphere. This is because condensation occurs with the utmost difficulty, if at all, in clean air. Moisture needs a suitable surface upon which to condense. Experiments have shown that if pure air is cooled beyond its dew-point it only becomes supersaturated. Condensation may not occur even though the relative humidity is in excess of 100%. On the other hand, if suitable surfaces on which moisture can condense exist in the air, condensation may start before the air is saturated. These surfaces can be land or plant surfaces as in the case of dew or frost. But in the free air condensation begins around hygroscopic nuclei. These are microscopic particles of substances like smoke, dust, sulphur dioxide, or salts (sodium chloride) which have the property of wettability. Condensation begins on these substances before the air is saturated and in the case of sodium chloride when the relative humidity of the air is about 78%.

The hygroscopic nuclei abound in the atmosphere and hence condensation easily takes place when the air is sufficiently cooled by any of the mechanisms earlier mentioned. These nuclei range in size from those with radius of 0.001 μm which are ineffective because the air must be highly supersaturated for condensation to occur to those with radius of over 10 μm which do not remain airborne for long (Barry and Chorley, 1976). The most effective nuclei for condensation purposes, therefore, are those lying between these two extremes. The dynamic process of adiabatic cooling is important for cloud formation and precipitation. It therefore deserves to be discussed in some detail.

Adiabatic temperature changes

When a parcel of air is vertically displaced for whatever reason certain changes occur in the air. Because the air parcel encounters lower pressure and there is generally no exchange of heat with the surrounding air, the volume of the vertically displaced air parcel increases following expansion. This process involves work and consumption of energy. So the heat available per unit volume of air decreases and there is a fall in temperature. Since such a temperature change does not involve energy gain from or loss to the environment, it is said to be adiabatic.

The rate at which temperature decreases in a rising and expanding air parcel is known as the *adiabatic lapse rate*. Until condensation occurs, temperature will fall at the rate of about 9.8 °C per kilometre. This is known as the *dry adiabatic lapse*. Continuous fall in temperature will lead to condensation when the dew-point temperature of the air parcel is reached. Latent heat will be released by the condensation process and this will slow down the rate of fall of temperature within the rising parcel of air. The air will now cool at a slower rate known as the *saturated adiabatic lapse rate*. The saturated adiabatic lapse rate is not constant like the dry adiabatic lapse rate. It varies with temperature. Because a hot air mass is able to hold more moisture than a cooler one more latent heat will be released on condensation. Hence, for high temperature the saturated adiabatic lapse rate is low (about 4 °C per kilometre) and increases with decreasing temperature reaching about 9 °C per kilometre when the temperature is – 40 °C.

In addition to the two dynamic lapse rates above, we have the static one. This is the *environmental lapse rate* which is the normal decrease of temperature with height within the troposphere (see Chapter 2). This is not an adiabatic lapse rate, the rate depending on the local air temperature conditions. The changing properties of moving air parcels can be expressed graphically on a tephigram. The tephigram can also be used to determine the following among others:

1 the level at which an air parcel becomes saturated if forced to rise, i.e. the condensation level;
2 the layering and stability or otherwise of the atmosphere at various levels (see Chapter 9).

Air stability and instability

An air parcel or air mass is said to be stable, neutral, or unstable if when subjected to some disturbing impulse, it returns to its original position, remains in its disturbed position, or moves farther from its original position respectively when the disturbing impulse is removed. In a *stable* air parcel or air mass, the environmental lapse rate is less than the dry adiabatic lapse rate. This means that if the air is forced to rise it will always be cooler and more dense than its surroundings and therefore will tend to revert to its original level. Also, if such air is forced to move downwards it will gain in temperature at the dry adiabatic lapse rate and will always be warmer and less dense than the surrounding air. It will therefore tend to return to its former position unless prevented from doing so for some reason.

In contrast when or where the environmental lapse rate is greater than the dry adiabatic lapse rate the air will always be warmer than the surroundings and therefore *unstable*. If the air parcel is forced downwards under this condition, it will always be colder than its surroundings. The characteristic of unstable air is its tendency to continue moving from its original position once set in motion.

There are two other types of instability condition known as *conditional* and *potential instability*. An air parcel may be stable in its lower layers and when forced to rise by convectional heating or orographic barrier it becomes warmer than its surroundings and rises up freely. This situation is what is known as *conditional instability* because the development of instability is dependent on the relative humidity of the air. The state of conditional instability occurs frequently since the environmental lapse rate is often between the dry and saturated adiabatic lapse rates. If after uplift, an air parcel becomes conditionally unstable such an air parcel is said to have been in a state of *convective or potential instability*.

Finally, an air parcel or air mass is said to be *neutral* if, when forced up or down, it has a tendency to remain in its disturbed position once the motivating force is removed. This situation occurs when the environmental and dry adiabatic lapse rates are equal. The determination of stability condition in an air parcel or air mass with the aid of tephigram analysis is discussed in Chapter 9.

Clouds

Clouds are aggregates of very minute water droplets, ice crystals, or a mixture of both with their bases well above the earth's surface. Clouds are formed mainly as a result of vertical motion of moist air as in convection, or in forced ascent over high ground or in the large-scale vertical motion associated with fronts and depressions.

Clouds are usually classified into types on the basis of two criteria. These are:

1 shape, structure, and form or appearance of the cloud;
2 the height at which the cloud occurs in the atmosphere.

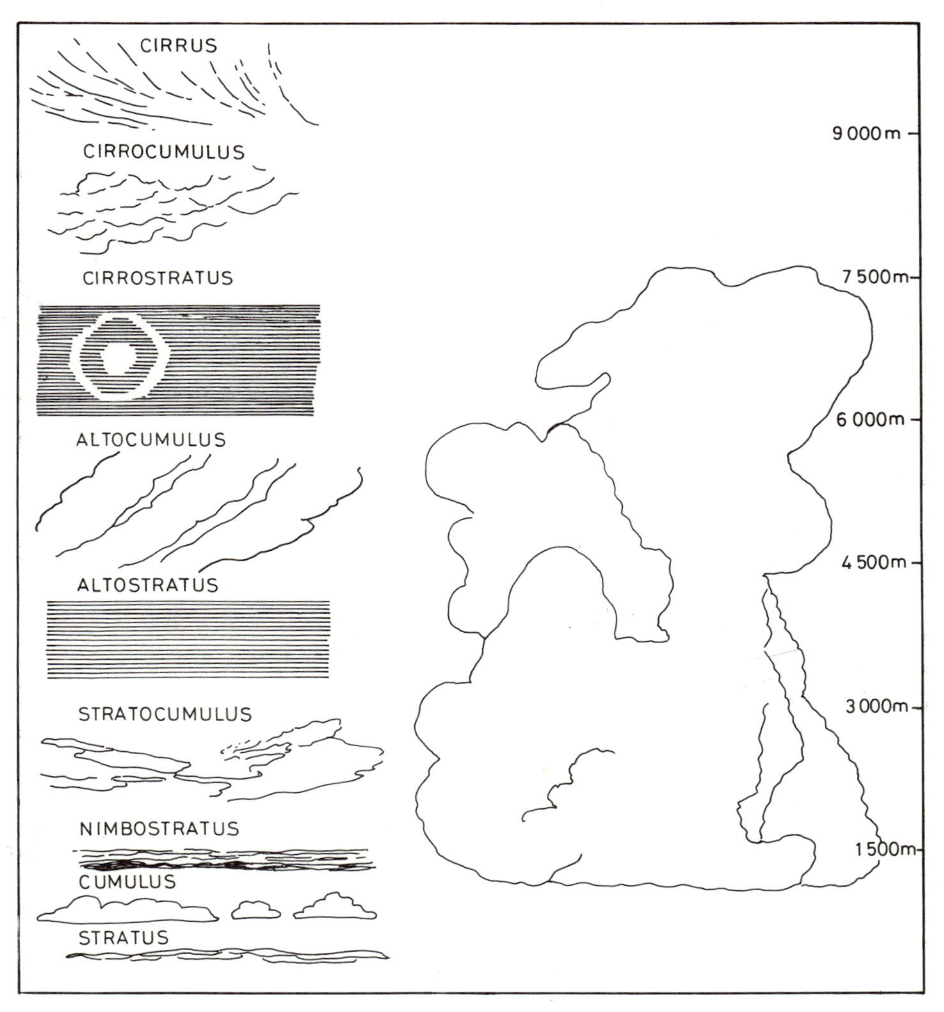

Fig. 7.5. Major types of clouds

Using the first criterion, we have the following major types of cloud:

1 cirriform clouds with fibrous appearance;
2 stratiform clouds which are in layers;
3 cumuliform clouds which have a heaped appearance.

Using the second criterion, we can identify low, medium, and high clouds. The height of the cloud base, however, varies with latitude as shown in Table 7.4. It also varies with the appearance or form of the cloud. For instance, cirriform clouds tend to be high clouds while cumulus clouds show such progressive vertical development that they cannot meaningfully be classified by the height of the cloud base. The ten basic cloud types which are recognized according to height and form are shown in Fig. 7.5 and Table 7.5.

It is also possible to classify clouds according to their mode of origin, i.e. the mechanism of vertical motion that produces condensation. Four broad such categories are as follows:

1 clouds produced by a gradual uplift of air in a depression;
2 clouds produced by thermal convection;

Table 7.4 Variations in height of the cloud base in various latitudinal zones (in metres) (after Barry and Chorley, 1976)

	Tropics	Middle latitudes	High latitudes
High cloud	Above 6000	Above 5000	Above 3000
Medium cloud	2000–7500	2000–7000	2000–4000
Low cloud	Below 2000	Below 2000	Below 2000

Table 7.5 Classification of clouds

Group	Mean upper and lower levels	Types of clouds
High clouds	6000–12,000 metres	Cirrus (ci) Cirrocumulus (Cc) Cirrostratus (Cs)
Medium clouds	2000–6000 metres	Altocumulus (Ac) Altostratus (As)
Low clouds	Ground level–2000 metres	Stratocumulus (Sc) Stratus (St) Nimbostratus (Ns) Cumulus (Cu)* Cumulonimbus (C_b)*

* These clouds may extend from above the earth's surface to a height of 6000 metres. Cumulus and cumulonimbus clouds are therefore often known as clouds with vertical development.

3 clouds produced by forced convection, i.e. mechanical turbulence;
4 clouds produced by ascent of air mass over a mountain barrier.

Each of these groups may, however, include a wide range of clouds of different structure and appearance. Besides, the same type of cloud may be produced by two different mechanisms of vertical motion. For instance, cumulus clouds may be produced by thermal convection or by the forced uplift of air over mountains.

Cloudiness or cloud amount is specified by the proportion of the sky covered by cloud of any type. This proportion is visually estimated in oktas. An okta is a unit of cloud amount measurement equal to area of one-eighth of the sky within the view of the observer. The latitudinal distribution of mean annual cloudiness over the earth indicates that cloudiness is lowest in the subtropics and highest in the high latitude (see Fig. 7.6).

The comparatively low value of cloudiness for the low latitudes is due primarily to the general absence of stratiform clouds there. The very low values in the subtropical belts are caused by the prevailing high pressure cells with subsiding air. The slightly higher values around the equator are associated with the low pressure and convergent air flow there.

Within a given zone, cloudiness varies with location and seasons. There are also diurnal variations. Seasonal variations are strong in the tropics particularly outside the equatorial belt. Over land, convectional processes cause an afternoon maximum of cloudiness in most parts of the tropics. As ground temperatures decrease at night, the air becomes stable and the amount of cloud cover decreases. Stratiform clouds may, however, persist through the night as they retard cooling during the night. On the other hand, over tropical water surfaces cloudiness shows a night maximum as a result of instability which is further intensified by radiation cooling from cloud tops.

The correlation between cloud amount and precipitation is not always a high or reliable one. If clouds are the stratiform type and/or are too thin little or no precipitation will be produced. The coastal areas of Namibia, Morocco, and Peru for instance have high amounts of cloud cover but these areas receive very little precipitation. Appreciable amounts of rain are usually obtained in the

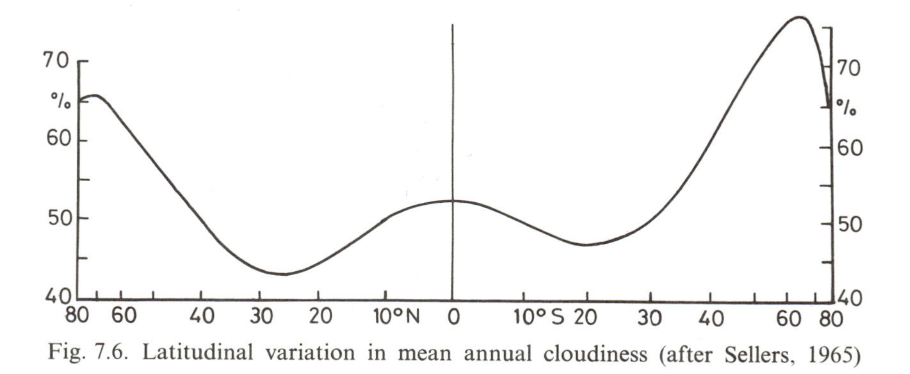

Fig. 7.6. Latitudinal variation in mean annual cloudiness (after Sellers, 1965)

tropics from cumulonimbus clouds while in the temperate region appreciable precipitation falls from nimbostratus cloud and in summer from cumulonimbus clouds. We may now examine the mechanisms by which precipitation is formed from the masses of microscopic water droplets and ice crystals that make the cloud.

Formation of precipitation

Various theories of raindrop formation have been put forward at various times and virtually all have been rejected for one reason or another. The two theories currently accepted are those which explain the growth of raindrops in terms of ice crystals growing at the expense of water droplets (the Bergeron–Findeisen theory) or in terms of the coalescence of small water droplets by collision and the sweeping action of falling drops (the coalescence theory).

According to the Bergeron–Findeisen theory of raindrop formation, ice crystals within clouds tend to grow larger at the expense of the water droplets until they become too heavy to be supported within the cloud and consequently fall. These ice crystals will melt to form raindrops if they encounter warmer air as they descend. If not, they will fall as snow. When the temperature near the earth's surface hovers around the freezing level, the ice crystals will partially melt and fall as sleet, a mixture of rain, and snow. The ice crystals grow larger at the expense of the water droplets because the saturation vapour pressure over ice is less than that over water. This means that vapour which is only saturated with respect to water is supersaturated with respect to ice. Condensation therefore occurs on the ice crystals at the expense of the supercooled water droplets.

The Bergeron–Findeisen theory is supported by observation and laboratory experiments. For instance, in extratropical areas it is generally observed that significant precipitation always comes from clouds whose tops extend beyond the freezing level in the atmosphere whereas lower clouds yield no more than mist or drizzle. Also, radar experiments have confirmed the existence of both water droplets and ice crystals in clouds extending beyond the freezing level and that such clouds give significant precipitation. Finally, in cloud seeding experiments clouds have been made to produce rain by seeding with fine dry ice (solid form of CO_2) or silver iodide which has a crystal structure very similar to that of ice.

The Bergeron–Findeisen theory cannot, however, be used to explain the process of raindrop formation in tropical clouds which are usually warm because they do not extend into the freezing level in the atmosphere. These clouds are therefore made up solely of water droplets. And yet these clouds produce heavy rainfall. Within such clouds, raindrops grow by the coalescence process. The larger water droplets within the clouds fall at a faster rate than the smaller ones, overtaking and absorbing the smaller droplets along their paths. The larger droplets also drag or sweep the smaller ones and absorb them (see Fig. 7.7). Experimental results indicate that the coalescence process allows a more rapid growth of raindrops than simple condensation although it is initially

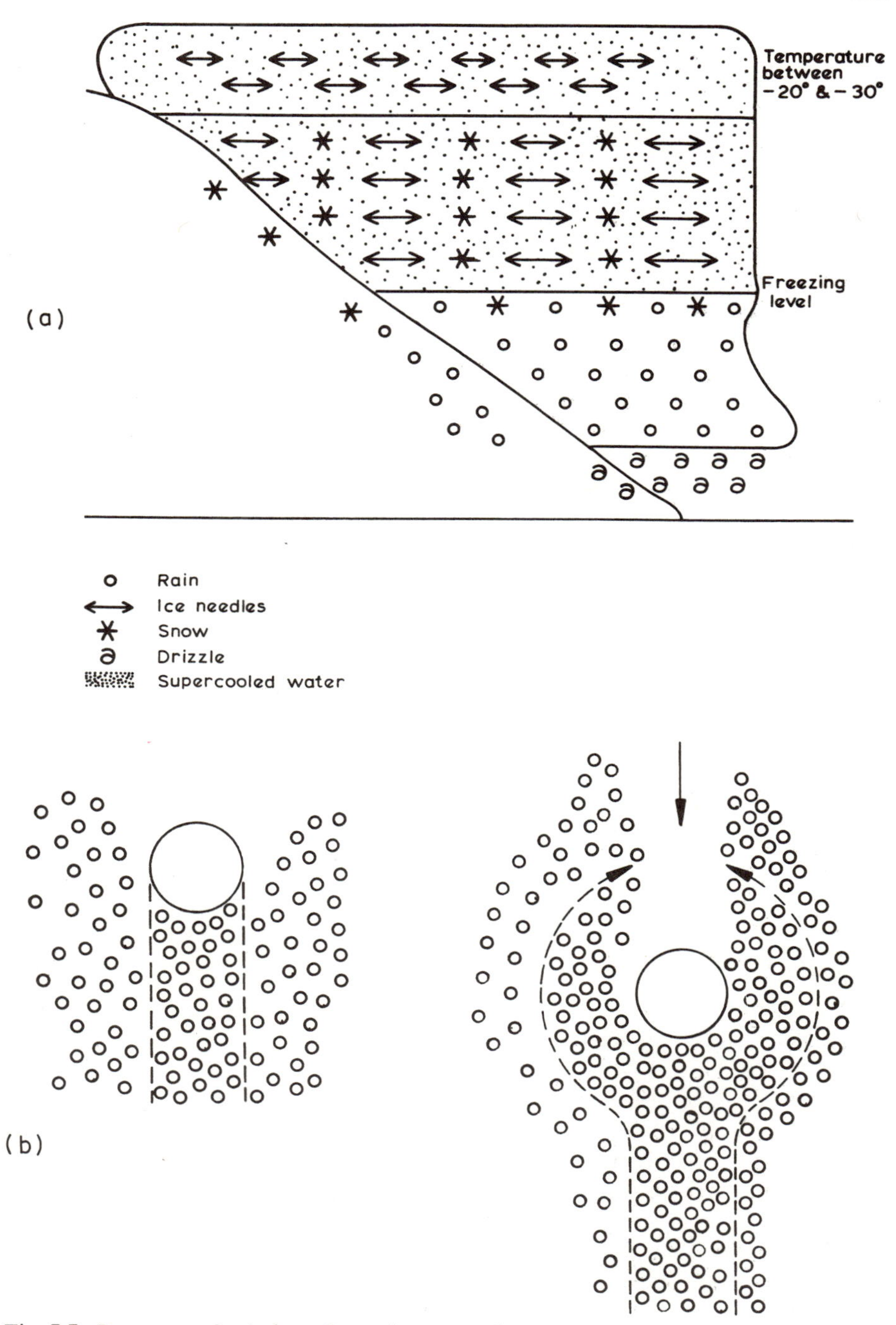

Fig. 7.7. Processes of raindrop formation according to: (a) Bergeron–Feideisen theory and (b) coalescence theory

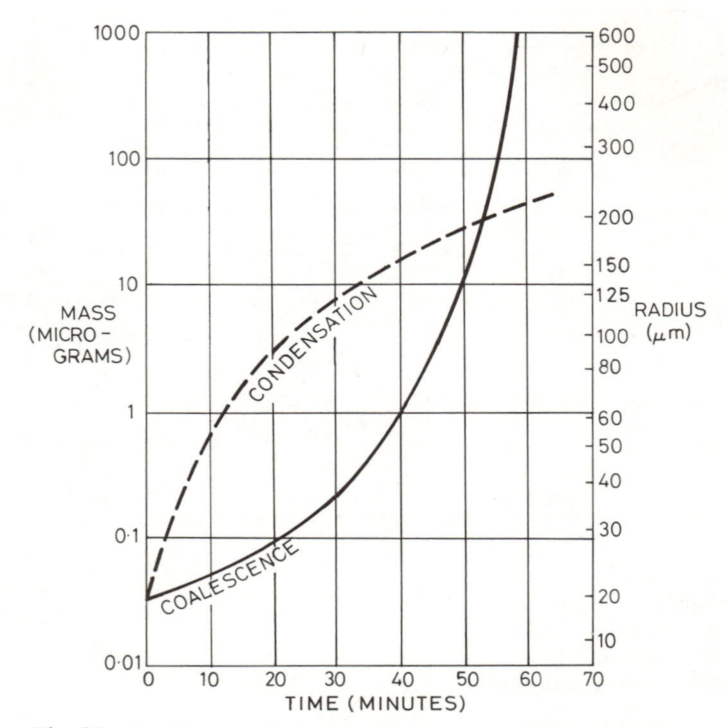

Fig. 7.8. Droplet growth by condensation and coalescence (from Barry and Chorley, 1976)

rather slow (see Fig. 7.8). The coalescence process has been found to occur in clouds in tropical maritime air masses in the temperate latitudes particularly in summer.

Other condensation phenomena

Apart from clouds and precipitation, other weather phenomena associated with condensation particularly at or near the earth's surface include dew, fog, and frost.

Dew is the condensation of water vapour on a surface whose temperature has been reduced by radiational cooling to below the dew point temperature of the air in contact with it. There are *two* processes of dew formation. First, under calm conditions water vapour diffuses upwards from the soil to the exposed cooling surface such as grass in contact with it and there condenses.

Second, under conditions of light wind, downward turbulent transfer of water vapour from the atmosphere to the cooled surface takes place. This is known as dewfall and occurs less commonly than the first process.

The two processes take place on clear relatively calm nights when radiational cooling is favoured. For dewfall to occur the air must in addition be moist and the wind light but not excessive.

The term 'fog' is used to describe obscurity in the surface layers of the atmosphere in which visibility is less than 1 km. The obscurity in the atmosphere may be due to minute water droplets or smoke particles or both suspended in the atmosphere.

Fogs may be entirely composed of smoke or dust particles although many fogs particularly in industrial areas are composed of both smoke particles and water droplets. Such fogs are also known as smog.

Fogs which are composed entirely or mainly of water droplets are classified according to the physical process that produces air saturation. Thus, we have the following types of fog:

1 radiation fog which forms on land on clear, calm nights with moist air;
2 advection fog which is formed when a relatively warm, moist, and stable air moves over cold water or land surface;
3 upslope or hill fog which is formed on windward slopes of hills by forced uplift of moist stable air until saturation is reached as a result of adiabatic cooling by expansion;
4 evaporation fog which is formed by the evaporation of relatively warm water into cool air. There are two types of this fog; the steam fog or arctic sea smoke which develops when cold air moves over a warm water surface so that the water appears to steam as the warm water is evaporated into the cold air and the frontal fog which occurs at or near the boundary between two air masses (front) as a result of rain falling from the relatively warm air above a frontal surface being evaporated into the cooler, drier air below causing it to become saturated.

It should be noted that fogs in nature frequently result from the combined action of two or more of the above physical processes.

Frost is said to occur when the temperature of the air in contact with the ground or at screen level is below the freezing point giving 'ground frost' and 'air frost' respectively. The term frost is also used to describe the icy deposits on the ground or on objects in such temperature conditions. The significance of frost in agriculture is discussed in Chapter 12.

References

Barry, R. G. and Chorley, R. J. (1976). *Atmosphere, Weather and Climate* (3rd edn). Methuen, London.

Bowen, I. S. (1926). The ratio of heat losses by conduction and by evaporation from any water surface. *Physical Review* **27**, 779–787.

Chang, Jen-hu (1964). On the study of evapotranspiration and the water balance. *Erdkunde* **19**, 141–150.

Garnier, B. J. (1956). A method of computing potential evapotranspiration. *Bulletin de L'Ifan* **18**, A 3, 665–676.

Gates, E. S. (1972). *Meteorology and Climatology for the Sixth form and Beyond* (4th edn). Harrap, London.

Landsberg, H. E. (1966). *Physical Climatology*. Gray Printing Co., Pennsylvania.

Penman, H. L. (1948). Natural evaporation from open water, bare soil and grass. *Proceedings Royal Society Series A* **193**, 120–145.

Sellers, W. D. (1965). *Physical Climatology*. University of Chicago Press, Chicago.

Sibbons, H. L. (1962). A contribution to the study of potential evapotranspiration. *Geografisker Annaler* **44**, 279–292.

Thornthwaite, C. W. (1948). An approach toward a rational classification of climate. *Geographical Review* **38**, 55–94.

Thornthwaite, C. W. and Holzman, B. (1939). The determination of evaporation from land and water surfaces. *Mon. Wea. Rev.* **67**, 4–11.

CHAPTER 8

Precipitation

Measurement of precipitation

In meteorology, the term 'precipitation' is used for any aqueous deposit in liquid or solid form derived from the atmosphere. Consequently the term refers to various liquid and frozen forms of water like rain, snow, hail, dew, hoar frost, fog-drip, and rime. However, only rain and snow make significant contributions to precipitation totals and in the tropics the term rainfall is interchangeable with precipitation since snow is generally absent except on some high mountains like the Kilimanjaro in East Africa. Snowfall is difficult to measure accurately and most precipitation records are in fact rainfall equivalent records. In this chapter therefore rainfall and precipitation are used interchangeably.

The depth of fresh snowfall is usually measured by a graduated ruler. The water content is measured by means of a snow gauge which is really the normal rain gauge fitted with some devices to collect solid precipitation and melt it before reading takes place. One metre of freshly fallen snow when melted gives the same amount of water as 10 cm of rainfall. Snowfall can therefore be measured in the absence of a snow gauge by collecting and melting samples of fresh snow which has fallen in the open.

The measurement of rainfall, though more accurate than that of snowfall, is itself beset with some problems. Rainfall is measured with the aid of the rain gauge of which there are two basic types, the recording (or autographic) and the non-recording rain gauge. In the former type of rain gauge there is a mechanism to measure the volume or weight of rainfall reaching the gauge as well as its timing. There are three main types of recording rain gauges depending on the main operating mechanisms which may be the tilting siphon, the tipping bucket, or the weighing collector system. Whatever the operating mechanism, a record is made on a moving chart and sometimes on punched or magnetic tape.

The tilting siphon autographic rain gauge is perhaps the most widely used autographic rain gauge. The gauge is a large version of the standard gauge but it contains a collecting chamber fitted with a float. As rain falls and the chamber fills up, the float rises and a pen attached to the top marks a line rising up a chart fixed to a cylindrical drum driven by clockwork. When the chamber is filled with enough rain it tilts over on its pivot and the contents siphon out of the gauge. The float returns to its original level and the pen now rests at the base of the chart. The chamber fills up again as the rain continues to fall and the whole process is repeated until the rain stops. One complete cycle measures 5 mm of rainfall. In heavy rainstorms the pen quickly performs several rises and falls. During dry periods a horizontal line is drawn round the base of the chart. The charts are usually changed every 24 hours. An autographic rain chart has horizontal lines representing the amount of rainfall in millimetres and vertical lines representing rainfall duration in hours and minutes. Using such a chart we can determine not only the volume of a given rainstorm but also its duration and average intensity. Rainfall intensity recorders which record the intensity of rainfall at any instant are now available. In the case of a non-recording rain gauge only the total volume of rainfall can be measured. Rainfall enters a collection can through an orifice of known area and the collected rainfall is measured using a measuring jar graduated in millimetres.

The amount of rain caught by a given rain gauge in a given location depends on a number of factors such as the height of the gauge above the ground, wind velocity, and rate of evaporation. Factors of site location are also important.

Table 8.1 Variations of rain catch with gauge height

Height of gauge mouth above the ground	Rain catch as % of that at 0.3 m (1 ft)
50 mm	105
100 mm	103
150 mm	102
200 mm	101
300 mm	100
0.45 m	99.2
0.75 m	97.7
1.5 m	95.0
6 m	90.0

Table 8.2 Effect of wind speed on rain gauge catch

Windspeed in metres per second	% Deficit from true catch
0	0
2	4
4	10
6	19
8	29
10	40
12	51

Experiments have shown that the higher the height of the gauge above the ground the less the amount of rain caught (see Table 8.1). Similarly, the amount of rain caught by a rain gauge tends to decrease with increasing wind velocity or turbulence (see Table 8.2). Rain gauges must therefore be well sited away from any obstructions like trees or walls capable of influencing the amount of rain caught.

Precipitation types

It is conventional to classify precipitation into three main types on the basis of the mode of uplift of air that has given rise to the precipitation. These types are:

1 convective type precipitation associated with convective instability;
2 cyclonic type precipitation associated with convergence in a depression; and
3 orographic precipitation associated with hilly or mountainous areas.

Each of these precipitation types will now be described.

Convective type precipitation is associated with cumulus and cumulonimbus clouds. Precipitation is caused by vertical motion of an ascending mass of air which is warmer than its environment. Convective type precipitation is usually more intense than cyclonic or orographic precipitation though it is normally shorter in duration. Convective type precipitation is often accompanied by thunder. Depending on the degree of spatial organization of the precipitation the following three subcategories of convective type precipitation can be identified.

1 Scattered convective showers lasting half to one hour may occur over an area of 20–50 square kilometres following intense solar heating of the land surface, particularly in summer. The precipitation is of the thunderstorm type and often includes hail.
2 Organized convective showers may form as a result of intense insolation over high grounds in the tropics or when moist unstable air passes over a warmer surface. Such convective cells move with the wind and occur parallel to a surface cold front or ahead of the warm front. Precipitation is widespread though it may be of short duration at a given locality.
3 Cumulonimbus clouds organized about the vortex of tropical cyclones bring prolonged and heavy precipitation over large areas.

Cyclonic precipitation is caused by the large-scale vertical motion of air associated with low pressure systems like the depressions. Precipitation is moderately heavy, continuous, and affects very extensive areas as the depression moves. Cyclonic precipitation is not as intense as convective type precipitation but it has longer duration. Often, cyclonic precipitation lasts for 6–12 hours.

Orographic precipitation is usually defined as precipitation that is caused entirely or mainly by the forced uplift of mosit air over high gound. Mountains *per se* are, however, not very efficient in causing moisture to be removed from the air mass moving across them. None the less, hilly areas receive more precipitation than their surrounding low lands. In addition windward sides of mountains are known to receive more precipitation than the leeward sides which are said to suffer from the rain shadow effect of the windward slopes. The degree of influence of mountains on precipitation depends on their size and their alignment relative to the rain bearing winds. It also depends on the stability or otherwise of the atmosphere as well as the moistness of the air mass. In a stable atmosphere orographic influence is restricted to the proximity of the mountain or hill so that the main effect of high ground is just to redistribute precipitation. On the other hand, when the atmosphere is unstable orography tends to increase the amount of precipitation as well as distribute it over a larger area. Mountains may influence precipitation in one or more of the following ways (Barry and Chorley, 1976).

1 They trigger conditional or convective instability by giving an initial upward motion to the air stream or by differential heating of the mountain slopes which are differently exposed to insolation.
2 They increase cyclonic precipitation by slowing down the rate of movement of depressions.
3 They cause convergence and uplift through the funnelling effects of valleys on air streams.
4 They encourage turbulent uplift of air through surface friction. Stratus or stratocumulus clouds may be formed under such a situation to give some drizzles or light precipitation.
5 Orography may also influence precipitation through frictional slowing down of an air stream moving inland from the sea. Convergence and uplift of air may be initiated in this manner.

Global distribution of precipitation

The distribution of precipitation over the earth is considerably more complex than that of insolation or air temperature (see Fig. 8.1). Since, as mentioned in Chapter 7, practically all precipitation results from adiabatic cooling of ascending air masses, rainfall is highest in areas of air mass ascent. The major ones are the zones of convergent horizontal air flow in the equatorial belt and the atmospheric disturbances of the middle latitudes as well as areas along the windward sides of mountain ranges.

The following are the major features of world pattern of precipitation as shown in Fig. 8.1.

1 There is abundant precipitation in the equatorial zone and moderate to large amounts in the middle latitudes.
2 The subtropical belts and the areas around the poles are relatively dry.
3 The west coasts in the subtropics tend to be dry while the east coasts tend to be wet.

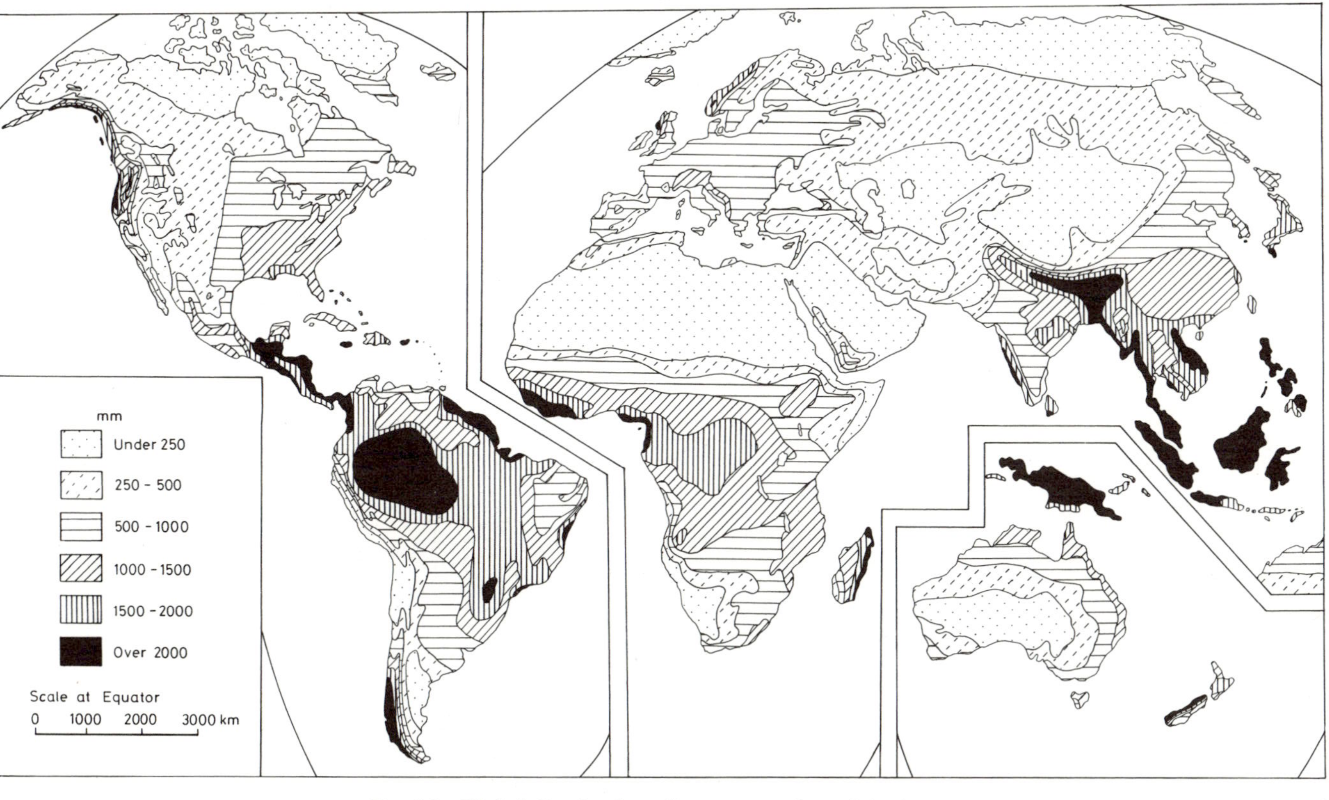

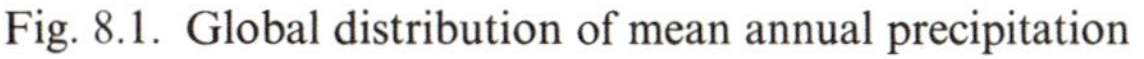
Fig. 8.1. Global distribution of mean annual precipitation

4 In the high latitudes it is the west coasts that are in general wetter than the east coasts.
5 Precipitation is abundant on the windward sides of mountains but sparse on the leeward sides.
6 Areas close to large water bodies like the oceans receive more precipitation than the interiors of the continents which are located far away from sources of moisture supply in the oceans.

The distribution of average annual precipitation by latitude zones is shown in Fig. 8.2. As shown in the graph, precipitation over ocean surfaces is higher than that over land. There is a strong maximum precipitation on both the land and the ocean around the equator where convergence of surface winds gives rise to a large-scale uplift of warm, humid, and unstable air. The amount of precipitation decreases poleward in each hemisphere, reaching secondary minima in the subtropics characterized by high pressure with divergent wind systems and vertical subsidence. Poleward from the subtropics, precipitation increases to reach a secondary maximum about latitudes 40–50° north and south of the equator. These are the zones of middle latitude convergence and the associated depressions. Poleward from about latitudes 50–55° in both hemispheres precipitation decreases rather sharply to reach primary polar minima beyond latitudes 75° north and south of the equator (Trewartha, 1968).

The distribution pattern of precipitation over the globe is therefore rather complex owing to the influence of several factors such as topography (elevation), distance from large water bodies, direction and character of the prevailing air

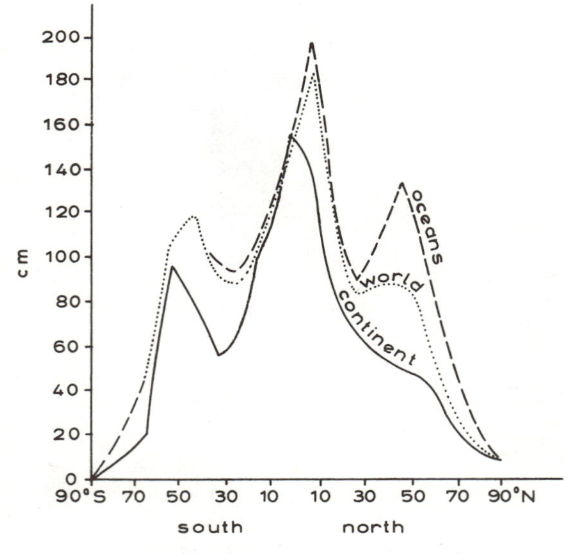

Fig. 8.2. Distribution of annual precipitation by latitudinal zones (after Trewartha, 1968, reproduced with permission)

masses among others. These factors can be grouped into two classes: those which influence the vertical motions of the atmosphere (e.g. convergence/divergence, atmospheric disturbances, and orographic barriers) and those which relate to the nature of the air itself, particularly its stability or instability and its thermal and moisture characteristics. These characteristics are determined by the nature of the source region of the air mass, its subsequent trajectory and age among others (see chapter 6).

Seasonal variations in precipitation

The seasonal distribution of precipitation is as important as the total amount both in the tropics and in extratropical areas. In many parts of the tropics precipitation occurs mainly during the summer half of the year and the winter half is relatively dry. The times of the start, duration, and end of the rainy season control agricultural activities in the tropics. The rainy season also brings lower temperatures and exercises considerable influence on the people's way

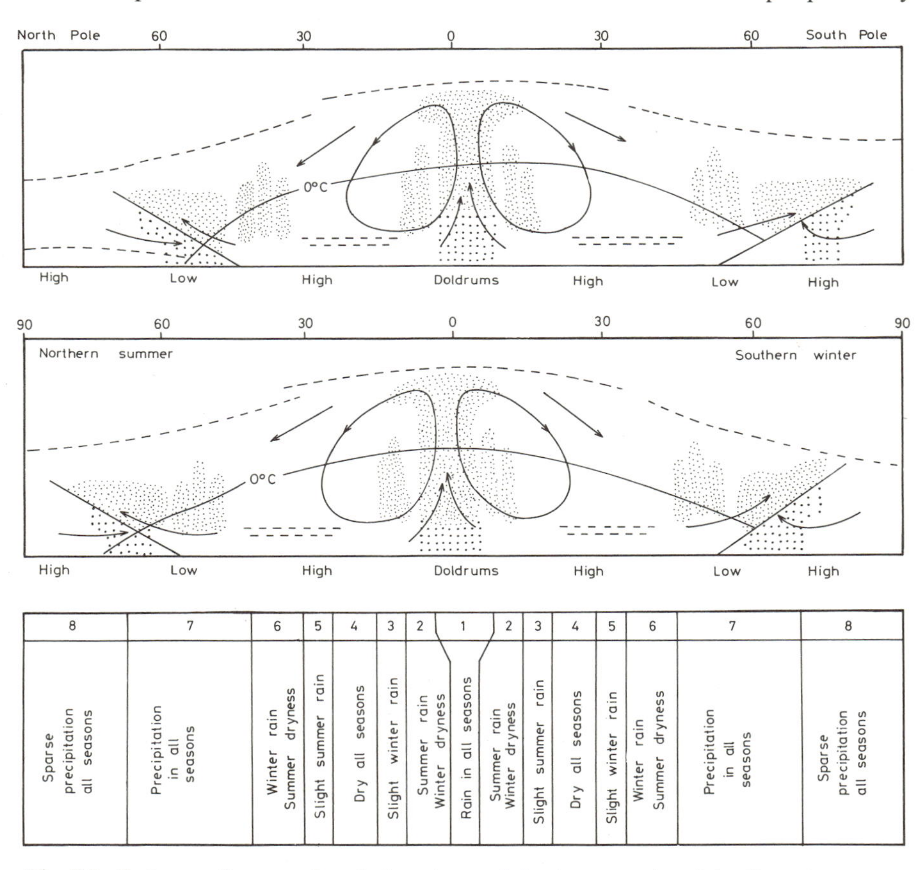

Fig. 8.3. Patterns of seasonal variations in precipitation over the globe (from Pettersen, 1969, reproduced with permission)

of life by limiting outdoor activities. Because temperature and other climatic elements are much more uniform, the seasonal rainfall distribution forms the basis of most classifications or subdivisions of tropical climates.

Seasonal distribution of precipitation is also an important element of weather and climate in the middle and high latitudes. Whereas in the tropics rainfall is effective for plant growth whatever time of the year it falls, in the middle latitudes only the part of the annual precipitation that falls during the freeze-free season may be effective. The precipitation in winter is mainly in the form of snow which cannot be utilized by plants until it melts. Besides, temperatures are often too low during this same season for plant growth.

Precipitation tends to be more seasonal in its incidence in the tropics compared to the extratropical areas. The seasonal march of precipitation in the low latitudes is controlled primarily by the north–south migration of the wind belt which together with their associated zones of convergence and divergence follow the course of the sun. Also, the distribution pattern of seasonal precipitation is more zonal in the low latitudes than in the middle latitudes. In the latter areas the continents and oceans exert considerable influence on the pattern of precipitation distribution. Finally, oceanic areas not only receive more precipitation over the year than the land areas but their precipitation is also less seasonal in its incidence. Fig. 8.3 shows the general pattern of the variations in seasonal precipitation over the globe. Emphasis in this schematic diagram is on the north–south migration of zones of convergence and divergence. Seasonal variations in precipitation arising from non-zonal factors like land and water distribution, arrangement of highlands and longitudinal variations in atmospheric circulation are not considered.

The following principal rainfall regimes (i.e. patterns of seasonal rainfall) are generally recognized:

1 equatorial rainfall—rainfall is abundant, occurs all the year round and is largely convectional in origin;
2 savanna rainfall—rainfall is largely convectional and occurs during summer;
3 tropical desert rainfall—low rainfall at all seasons;
4 mediterranean rainfall—rainfall is mainly cyclonic (i.e. frontal) and occurs in winter. Summer is dry;
5 west European rainfall—precipitation is abundant with more falling in winter than in summer. Precipitation is mainly cyclonic in origin;
6 continental rainfall—rain falls mainly in summer;
7 east coast rainfall—rainfall is high coming from onshore winds in low latitudes; in the middle latitudes the precipitation is derived from inflowing warm moist air mass in summer and cyclonic storms in winter;
8 polar rainfall—precipitation is low with maximum precipitation occurring in summer when there is more moisture in the air and the cyclonic influence can reach the area around the poles.

Examples of some of the above various rainfall regimes are portrayed in Figs. 11.3–11.13.

Diurnal variations in precipitation

There are notable variations in precipitation incidence in the course of a day. Although the diurnal rainfall regime is not as important as the seasonal rainfall regime it is nevertheless an important feature of climate particularly in the low latitudes. Diurnal rainfall regime has a strong effect on air traffic and other means of communications particularly roads where they are unpaved. It also controls many outdoor activities and influences the efficiency of rain for agricultural purposes. Rain that falls during the day is subject to heavy evaporation losses particularly in the low latitudes.

Diurnal variations in precipitation in the middle latitudes are not as regular as those in the low latitudes and are therefore rather unpredictable. This is due to the difference in the dominant rain mechanism in the two areas. The convectional processes dominant in the tropics are strongly diurnal in character whereas in the middle latitudes the unpredictable disturbances and fronts predominate. Few studies of the diurnal variation in precipitation exist. In the middle latitudes diurnal variations in precipitation have attracted only limited interest because of their irregular and unpredictable pattern. In the tropics where diurnal rainfall regimes are more regular their studies have been hampered by lack of suitable data. Diurnal rainfall regimes can only be indicated by hourly rainfall data obtainable from autographic rain gauges which are still limited in most areas of the tropics.

Although diurnal variation in precipitation is a rather complicated phenomenon, two general types of diurnal rainfall regimes can be recognized. These are:

1 the continental or inland type; and
2 the marine or coastal type.

The continental type is found over large land masses (continents) where more of the rainfall tends to occur in the warmer hours of the day when solar heating is most intense and the lapse rate is steepened so that the air becomes buoyant. Thus, the continental regime is characterized by a maximum of rainfall during the late morning or afternoon, and the rainfall is largely convectional in nature.

The marine regime is characterized by rainfall maximum during the night and the early hours of the morning when the maritime air is most unstable. This regime occurs over oceans and in coastal areas or near large lakes. It is thought to be caused by some kind of night-time convection that results from the steepening of the lapse rate as the upper troposphere is cooled by radiation losses from cloud tops (Nieuwolt, 1977).

The above two types of diurnal rainfall regimes are rarely found anywhere in their ideal forms. Most locations show some slight deviations from these ideal patterns owing to the effect of local factors on small-scale atmospheric processes involved in diurnal rainfall distribution. Such factors include topography, vegetation type, drainage conditions, albedo, form of the coastline, and the presence of swamps, lakes, rivers, or irrigated fields. All these may influence the time of rainfall in a given locality. Despite variations most inland stations show a tendency towards rainfall minimum during the morning and a maximum later in the day irrespective of the seasons. Deviations from the ideal marine regime type are larger at coastal locations. In fact, some people have questioned the basis for classifying diurnal rainfall regimes into only two types since very few stations can be classified into either. Three diurnal rainfall regimes have, for instance, been recognized in Nigeria. These are:

1 the afternoon rainfall regime of the Niger delta and the Jos Plateau;
2 the morning rainfall regime of the southwest coastlands; and
3 the diffused evening, night, and early morning rainfall regime covering over 75% of the country.

This implies that over most of the country the diurnal rainfall regime is neither continental nor marine (see Ilesanmi, 1972).

Thompson (1957) has also noted that the continental regime is not widespread at inland stations in East Africa. It seems therefore that few stations show the ideal continental or marine diurnal rainfall regime. The observed patterns are far more complex than these two theoretical regimes suggest.

Variability of rainfall

The long term mean precipitation amounts for a month, season, or year hardly indicate the regularity or reliability with which given amounts of rainfall can be expected. This is particularly the case in the low latitudes and in relatively dry areas where rainfall is known to be highly variable in its incidence particularly from one year to another.

The variability of rainfall is an important consideration in the tropics where rainfall not only tends to be more variable than in the temperate region but is also more seasonal in its incidence within the year. The less variable rainfall is the more reliable it is. This is because the index of variability is a measure of the degree of likelihood of the mean amount being repeated each year, season, or month depending on the period under consideration. There are various measures of variability but the two measures commonly used in precipitation studies are the *relative variability* and *the coefficient of variation.* The latter measure is the more efficient provided the data is normally distributed so that the relative variability index is used in situations where the data is not normally distributed. In general annual precipitation totals are normally distributed except in areas where the mean annual rainfall is less than 750 mm. Monthly and seasonal rainfall totals tend to be non-normal and unless the data are transformed to make them normal the relative variability index is usually used to show variability.

The variability of annual precipitation over the earth is shown in Fig. 8.4. There is an inverse relationship between rainfall amount and rainfall variability. Annual precipitation is most variable in dry and subhumid areas. Notable examples are the deserts and steppes of the tropics and temperate region and

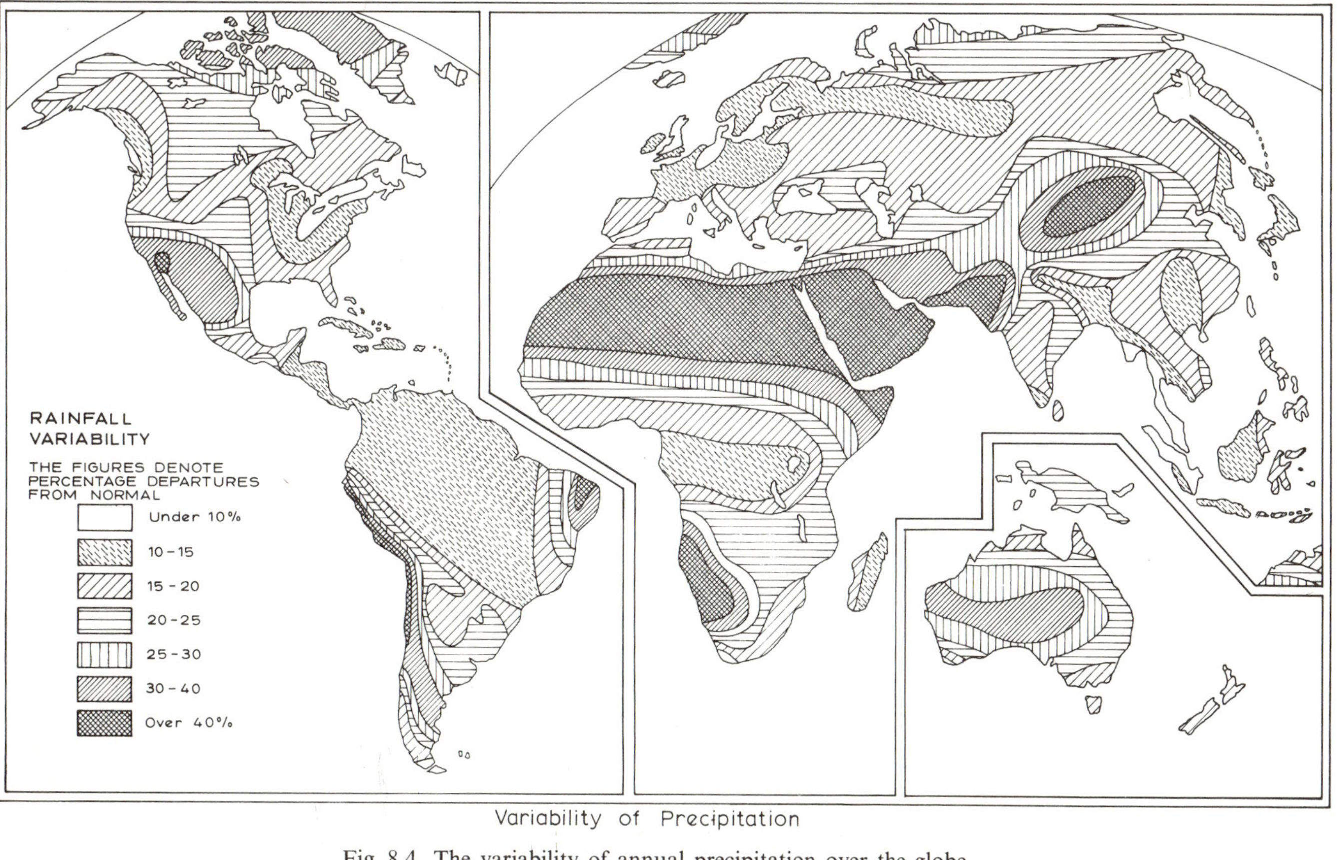

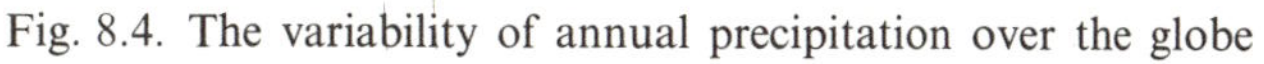

Fig. 8.4. The variability of annual precipitation over the globe

the cold lands of the high latitudes. Rainfall variability is small in humid regions of the wet tropics and cyclonic middle latitudes. The economic consequences of rainfall variability are, however, more serious in heavily populated agricultural areas of the humid regions than in sparsely settled dry or cold regions. The subhumid areas are the most vulnerable since small negative deviations from the mean rainfall may cause widespread crop failure and famine as frequently happens in the subhumid tropics. (See chapter 12 for a discussier on drought.)

Low variability implies that the mean precipitation at a given location is reliable while high variability implies wide fluctuations about the mean value. Rainfall reliability may, however, be defined in another way quite independent of the variability of rainfall. A given amount of precipitation may be regarded as reliable if it can be expected to be equalled or exceeded at some chosen probability level. In this regard, Manning (1956) has defined as reliable a rainfall amount that can be expected to be equalled or exceeded with 95% probability while Gregory (1964) is in favour of the 90% probability level. Alternatively, we may wish to compute the percentage probability that some critical rainfall amount will be exceeded or will not be reached. In the latter case, the procedure is to calculate the z-score (z) using the formula

$$z = \frac{x - \bar{x}}{\sigma} \tag{8.1}$$

where x is the critical rainfall amount and $\bar{x}$ and σ are the mean and standard deviation of the rainfall series respectively. The percentage probability that the rainfall will be more or less than the z-score corresponding to the critical value is then obtained from the *Table of the Normal Distribution Function.*

For the first case in which we are interested in the rainfall value that can be expected to be equalled or exceeded with a given probability we use the formula

$$x = z\sigma + \bar{x} \tag{8.2}$$

where x is the value which can be expected to be equalled or exceeded, $\bar{x}$ is the mean of observations, σ is the standard deviation and z is the z-score corresponding to the specified degree of probability and is determined from the *Table of Normal Frequency Distribution.*

Rainfall probability estimates and maps of the type described above have been used in determining the suitability of climatically marginal areas of East Africa for the growth of certain crops (see Manning, 1956).

Precipitation intensity

Precipitation not only varies in amount from one year, season, or month to another, it may also show a downward or upward trend over a given period. Rainfall fluctuations and trends are of great interest to climatologists as well as agriculturists in view of the important role of moisture in agriculture. Rainfall fluctuations and trends are discussed later in Chapter 10. Other precipitation characteristics such as duration, frequency, amount, and intensity of rainstorms are of hydrological interest. The intensity of precipitation during a given rainfall event is of vital interest to hydrologists and soil scientists concerned with flood forecasting and prevention and soil erosion control respectively. Precipitation intensity is the precipitation amount divided by storm duration in hours or minutes. Hyetograms from autographic rain gauges are required for the assessment of precipitation intensity. As mentioned earlier, rainfall intensity recorders for measuring instantaneous rainfall intensity are now available. Precipitation intensity varies with the time interval used. Average intensities for short periods are generally much greater than those for longer time intervals. Convective precipitation generally has more intensity than cyclonic precipitation. It is therefore not surprising that rainfall intensities are much higher in the tropics where convective rainstorms occur more frequently than in the middle and upper latitudes characterized mainly by cyclonic precipitation. High precipitation intensity has also been shown to be associated with increased raindrop sizes rather than an increase in the number of raindrops. The most frequent raindrop diameters for precipitation intensities of 0.1, 1.3, and 10.2 cm/h are 0.1, 0.2, and 0.3 cm respectively.

Where autographic rain gauges or rainfall intensity recorders are not available some idea of precipitation intensity is given by the amount of rain per rain day. This index, known as the *mean intensity*, may be computed for single months or for the whole year. A rain day is a period of 24 hours with at least 0.25 mm of rainfall. In other countries outside the Commonwealth critical values of 1, 2, and 5 mm have been used in defining a rain day. The values of the annual mean rainfall intensity at some stations mostly in the tropics are shown in Table 8.3. The mean precipitation intensity shows wide seasonal and spatial variations particularly in the tropics. The mean intensity generally increases as the total amount of precipitation increases except at high elevations of over 1500 metres where there is a decrease in the amount of rain per rain day as a result of the number of rain days increasing and the total precipitation decreasing (Nieuwolt, 1977).

Table 8.3 Annual mean rainfall per rain day in mm (rain days defined as days with at least 1 mm of rain)

	Rainfall (mm)		Rainfall (mm)
Quito, Ecuador	8.5	Bombay, India	22.4
Georgetown, Guyana	13.3	Calcutta, India	15.5
S. Juan, Puerto Rico	10.1	Rangoon, Burma	20.9
Accra, Ghana	13.6	Jakarta, Indonesia	13.5
Lagos, Nigeria	14.4	London, England	5.5
Entebbe, Uganda	12.4	Vienna, Austria	4.1

Rainstorms

Studies of the areal extent of rainstorms so far carried out show that pre-precipitation totals received in a given time interval vary according to the size of the area considered such that the smaller the area the more the precipitation received per unit area. Rainstorms constitute an important feature of tropical

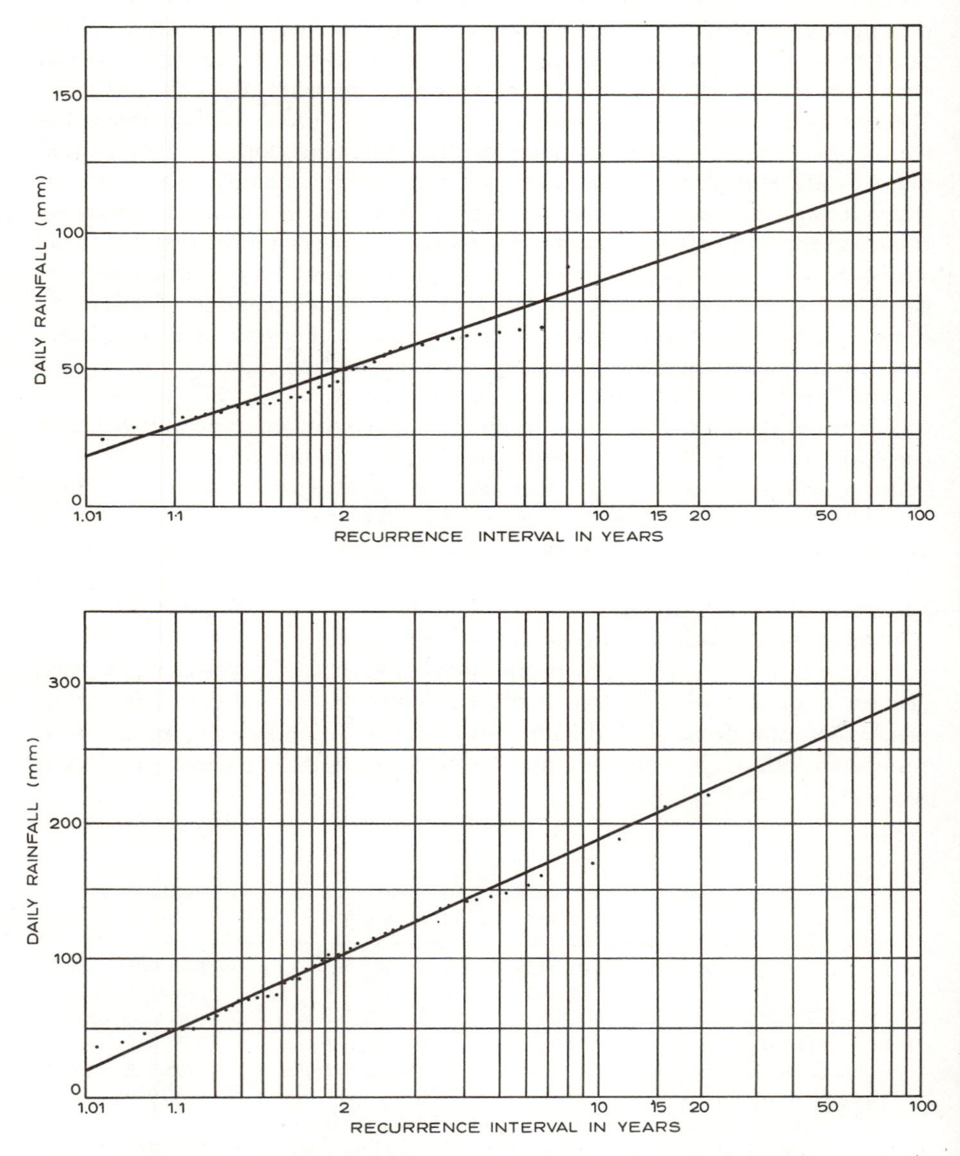

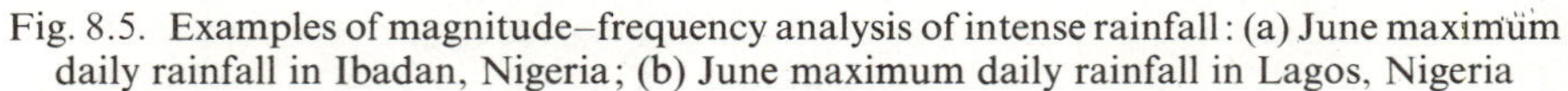

Fig. 8.5. Examples of magnitude–frequency analysis of intense rainfall: (a) June maximum daily rainfall in Ibadan, Nigeria; (b) June maximum daily rainfall in Lagos, Nigeria

climate. They account for most of the total rainfall in many parts of the tropics and account for the high intensities of tropical rainfall. Rainstorms are largely convective in origin and develop only during the warm season in extratropical areas. Rainstorms tend to produce a spotty rainfall distribution pattern for any given day or even as long a period as a month. When long term means are mapped, however, the spottiness is obscured because rainstorms generally occur in a random pattern. A very dense network of autographic rain gauges is required in studies of tropical rainstorms although their seasonal and regional variations can be fairly deduced from daily rainfall records (Nieuwolt, 1977).

Another precipitation characteristic which is of interest particularly to the hydrologist is the frequency of heavy rainstorms. Hydrologists are interested in the average time period within which a rainfall of specified amount or intensity can be expected to occur once. This is known as the recurrence interval or return period. Such information is required for a variety of design purposes (e.g. bridges, culverts, dams, and other hydraulic structures). A special type of graph paper known as the Gumbel extreme probability paper is often used in estimating rainfall amounts for given return periods or the return periods for specified rainfall amounts (see Fig. 8.5).

First the observed rainfall values are ranked in descending order. Second, the recurrence interval of each observation is computed using the formula

$$T = \frac{n+1}{m} \tag{8.3}$$

where T is the recurrence interval in years, n is the number of observations in the series and m is the rank of a particular observation. Third, the observed rainfall values are plotted against their computed recurrence intervals. Finally, a continuous line is fitted to the scatter of points by eye.

If the observed rainfall series does not conform to the Gumbel theory of extremes the scatter of points will not be linear. Other methods such as the normal probability theory will then have to be used. The same procedure outlined above is adopted for plotting of rainfall data on the normal probability paper.

References

Barry, R. G. and Chorley, R. J. (1976). *Atmosphere, Weather and Climate* (3rd edn). Methuen, London.

Gregory, S. (1964). Annual, seasonal and monthly rainfall over Mozambique. In Steel, R. W. and Prothern, R. M. (eds.) *Geographers and the Tropics*. Longman, London.

Ilesanmi, O. O. (1972). Diurnal variation of rainfall in Nigeria. *Nigerian Geographical Journal* **15**, No. 1, 25–34.

Manning, H. L. (1956). The statistical assessment of rainfall probability and its application to Uganda agriculture. *Proceedings Royal Society Series B* **144**, 460–480.

Nieuwolt, S. (1977). *Tropical Climatology*. John Wiley, London.

Thompson, S. W. (1957). The diurnal variation of precipitation in East Africa, *East African Meteorological Department Technical Memorandum*. No. 8, 70 pp.

Trewartha, G. T. (1968). *Introduction to Climate* (4th edn). McGraw-Hill, New York.

CHAPTER 9

Weather Observation, Analysis, and Forecasting

Weather observations

Measurements of weather elements or weather observations are carried out at locations known as the weather stations. Four types of weather station can be recognized depending on the number of weather elements measured, the frequency of measurement, and the status of the observer weather professional or amateur. The four types of weather station are as follows:

1 *synoptic stations*—these are stations manned by full-time professional observers who maintain continuous weather watch and make hourly instrumental observations of the weather elements on which information is required for the compliation of the synoptic charts or weather maps used in weather forecasting.
2 *agricultural stations*—these are stations manned by part-time observers making at least twice daily instrumental observations of the major weather elements. Evaporation, grass and soil temperatures, and solar radiation are also usually measured in view of their obvious importance in agriculture.
3 *climatological stations*—these are stations manned by part-time observers making only once or twice daily instrumental observations of temperature, humidity, rainfall, and wind.
4 *rainfall stations*—these are stations manned by part-time observers who take daily readings of rainfall only.

At a synoptic station, observations are made at fixed observing hours. The main synoptic hours internationally agreed upon are 0000 (midnight), 0600 (6 a.m.), 1200 (noon) and 1800 (6 p.m.) Greenwich Mean Time. Additional observations are made at other times between the four main times, often hourly or at three-hour intervals. The elements observed are as follows:

1 present and past weather—visually observed;
2 wind direction and speed—observed using the wind vane for direction and the cup anemometer for the speed;

3 amount and form of cloud—visually observed;
4 height of cloud—estimated using cloud height searchlights;
5 visibility—visually observed with the aid of objects spaced at known intervals or with the aid of visibility meter;
6 air temperature—measured with the aid of a thermometer placed in a Stevenson screen;
7 air humidity—measured with the aid of wet and dry bulb thermometers placed in a Stevenson screen;
8 barometric pressure—measured with the aid of a barometer;
9 precipitation—measured with the aid of a rain gauge;
10 sunshine duration—measured with the aid of Campbell–Stokes sunshine recorder.

The World Meteorological Organization (WMO) has recommended that in the design of a national network of stations the synoptic stations which are the principal weather stations should be spaced at intervals not exceeding 150 km. To ensure that observations at the different weather stations are accurate and comparable, the exposures of meteorological instruments should be similar. To this end, a weather station irrespective of the type should be located on a level ground covered with short grass and measuring at least 9 m by 6 m in size. The station should not be sited on or close to a hill, in a depression or on a steep slope. Likewise, it should be far from any obstacles like buildings or trees. Weather observation is a painstaking job that requires care, patience, honesty, and punctuality on the part of the observer. Weather observations must not only be accurate but must also be done on time.

Weather observations are made using various platforms. Apart from earth-based instruments at conventional weather stations we have airborne instruments designed mainly to probe the upper atmosphere and furnish information on the variations in the weather elements in the vertical over a given area. Such instruments may be carried aboard rockets, aircraft, helicopters or balloons. Since the early 1960s a dramatic improvement in weather observation has been achieved with the introduction of meteorological or weather satellites which among others give objective large area coverage of weather systems and are, therefore, useful in weather forecasting.

Weather observations fall into two main categories: those that are instrumental and those which are non-instrumental and depend on the skill, training, and judgement of the observer as they are carried out by eye observation. Weather elements like cloud type, cloud amount, and visibility are visually observed by observers with the requisite training and experience. The following meteorological instruments should be found at a standard weather station:

1 one or two Stevenson screens containing wet and dry bulb thermometers, maximum and minimum thermometer, thermograph, an hygrograph, barograph, or aneroid barometer;
2 an autographic rain gauge;
3 an ordinary rain gauge;

4 evaporation pans usually Class A Pans;
5 a wind vane and cup anemometer;
6 a sunshine recorder;
7 a radiation-measuring equipment such as the Gunn–Bellani radiation integrator or the Eppley pyheliometer.

Non-instrumental observations

Cloudiness and cloud forms are most commonly observed by the eye despite recent attempts at photographic recordings at a few places. The amount of cloud irrespective of height or type is visually observed and quantified in eights (Oktas) of the area of the sky that is covered with clouds. Thus 0 means a cloudless sky while 8 means the sky is completely covered with clouds. Cloud cover estimates are rather inaccurate, particularly at night, when only the presence or absence of stars is the guide to estimate amount of cloud cover. The forms of cloud are also visually observed, the appearance of the cloud being a major guide. The height of the cloud is assessed from its base and there are three categories—low, medium, and high (see Table 7.5 and Fig. 7.5). The types of precipitation are visually observed; the types include drizzle, hail, snow, sleet, etc. Visibility is assessed according to how far the observer can see into the distance. Usually markers are installed at known distances from the point of observation so that the distance away of the farthest marker the observer can see at a given time will indicate the degree of visibility at that time.

Instrumental observations

Most weather elements are instrumentally observed. The instruments are generally of two types—registering and non-registering types. Non-registering instruments have to be read at given times while registering types have some self-recording devices which give continuous values of the weather elements concerned in the form of a graph. The self-recording instruments vary in their degree of complexity and sophistication but are generally more expensive to purchase and maintain than the non-recording instruments. Nowadays, there are unmanned weather stations which can function perfectly for up to three months. Such stations are equipped with weather instruments which are fully automatic. Weather elements are continuously observed and the values recorded on electrical chart recorders or digital punched tape recorders for easy analysis using the computer. Automatic weather stations are very useful in difficult environments or in sparsely populated or remote areas. The only problem is that these instruments are expensive and being self-recording and automatic, require electricity or some other power source to function. For these reasons, automatic weather stations are at present virtually restricted to the advanced countries of Europe and North America.

Surface weather observations take place at conventional land-based weather stations and on the few weather ships that are located on the Atlantic and Pacific oceans. Commercial vessels also make some observations and radio these to nearby land-based central weather stations. There is a global network of over 7000 land-based weather stations and over 4000 merchant and passenger ships that make observations at sea. The network, particularly on land, is extremely uneven in coverage. The developing countries in the low latitudes and the polar areas are poorly served by weather stations. Similarly, the ocean areas do not have enough stations.

In addition to surface weather observations upper air observations also take place at many land stations and on some ships, particularly the weather ships which occupy fixed positions on the sea. This is done with the aid of radiosonde, radar, or electronic storm detecting (sferic) equipment. A *radiosonde* is a small radio transmitter equipped with instruments to measure pressure, temperature, relative humidity, wind speed and direction as it ascends into the atmosphere with the aid of an hydrogen-filled balloon (see Fig. 9.1). Readings are automatically transmitted and picked up by a radio receiver at an appropriately equipped ground station. When the balloon bursts the radiosonde falls to the earth suspended from a parachute. The height attained by the radiosonde

Fig. 9.1. A radiosonde being released into the atmosphere with the aid of a balloon

before falling is of the order of 16–20 km. Measurements above this height can only be made with the aid of rockets which can reach heights of 65–70 km.

Radar is basically a system of detection and location of targets which are capable of reflecting high frequency radio waves (microwaves). Information is presented visually on a cathode-ray tube. Radar is used in meteorology in the measurement of upper air winds and in the detection of storms, cloud, and precipitation elements. Information from weather radar is useful for routine weather forecasting particularly of the movement of rainstorms as well as for research into cloud physics and dynamics. The echo patterns from clouds are displayed on the radar screen in either of the following two ways:

1 the plan position indicator (PPI) which gives the positions of echoes in an almost horizontal plane surrounding the radar set; and
2 the range height indicator which scans in a vertical plane to produce a cross-section through the atmosphere so that the heights of bases and tops of echoes can be estimated.

The electronic storm detecting (sferic) equipment helps in pinpointing locations of thunderstorms. The electromagnetic waves caused by lightning flashes in the clouds are picked up at widely spaced stations having the equipment. By plotting the sferic fixes from a number of stations it is possible to pinpoint locations of thunderstorms with considerable accuracy.

A dramatic breakthrough in the observation of weather phenomena occurred in 1960 with the launching of the first *weather satellite* (see Chapter 1). Weather satellites provide two main types of vital meteorological information. First, they provide photographs of weather systems made up of various cloud forms on a regular and continuous basis which allows us to know the direction and rate of movement of these weather systems. Second, they measure radiation at the top of the atmosphere. Both the cloud photographs and radiation measurements from weather satellites have been found useful in general meteorological research as well as in weather and climate forecasting.

Weather satellite photographs are analysed in two ways: first, nephanalyses of the photographs can be made. These are maps or charts showing the distribution of clouds of various types over a large area. The predominant cloud types as viewed by the satellites as well as the orientation of medium- and large-scale cloud bands are represented on maps in the form of symbols. Such cloud charts can be interpreted in terms of atmospheric pressure patterns as weather systems. Second, satellite photographs may also be directly interpreted by taking and viewing them in sequence. Six main features of clouds are used for their identification: they are brightness, pattern, structure, texture, shape, and size. Cloud analysis, however, goes beyond recognizing cloud types. Significant synoptic features expressed in the cloud forms have to be identified. Most of these features are associated with the cyclonic vortex, hurricanes, or jet streams. The interpretation of satellite photographs requires skill and patience and a great deal of knowledge of meteorological theory and principles (Barrett, 1974).

Weather analysis

Weather can only be analysed after it has been observed. Before analysis is made, therefore, observations from both surface and upper air networks of stations must be assembled in a suitable form. The weather analyst is usually also the weather forecaster. To aid quick assemblage of weather information from places far and near a meteorological communications network has been set up throughout the world under the auspices of the World Meteorological Organization. Weather observations are first translated into an internationally agreed numerical code and then transmitted to various national meteorological centres by teleprinter, the signals of which are transmitted by radio. The forecaster at any given national meteorological centre can receive weather information from a large area often over a whole hemisphere and in time to be of use in forecasting. The amount of information received at a major national meteorological centre like Bracknell, in England, is very large and continues to grow so that methods of decoding and analysing this information are now largely automated.

The first job for a weather analyst is to construct an adequate three-dimensional picture of the atmosphere from available data which are never ideal or enough. Besides, changes are constantly taking place in the atmosphere both in the vertical and in the horizontal. It is therefore necessary to construct the three-dimensional picture of the atmosphere at frequent and regular intervals. Weather charts of the various weather elements are constructed for different selected levels of the atmosphere. Since it is impossible to construct such charts for all levels in the atmosphere, the basic surface and upper air charts are usually supplemented by cross-section analyses, upper air soundings and, in recent years, imagery from radar and the weather satellites as mentioned earlier.

Weather charts and diagrams are constructed according to agreed procedures laid down by the World Meteorological Organization. Surface and upper air observations are plotted on standard base maps by means of letters, figures, and symbols arranged around the points of observations in fixed positions. The standard plotting model for a station is shown in Fig. 9.2. The position of

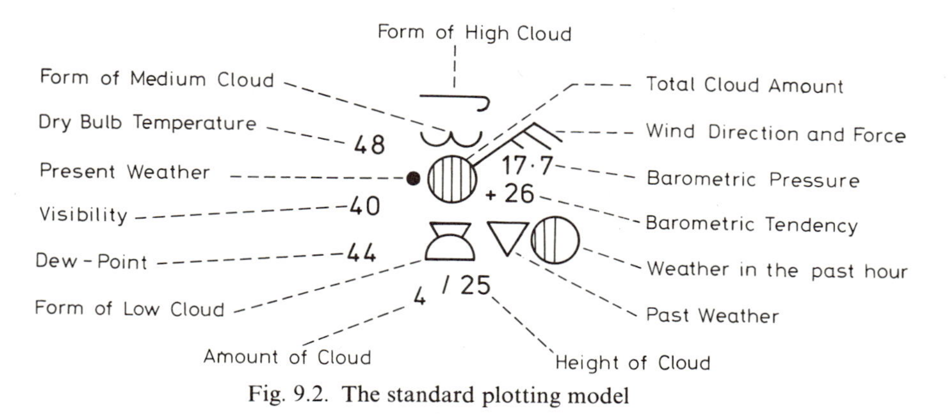

Fig. 9.2. The standard plotting model

PRESENT WEATHER

APPEARANCE OF SKY – no symbols used: b – blue sky (< ¼ covered): bc – ¼ to ¾ covered: c – cloudy (> ¾ covered) but with openings: o – overcast (continuous sheet): g – gloom

	Thinning last hour	No change	Thickening last hour
FOG (f) Fog, sky discernible			
In patches. Thick fog.			

HAZE (z) ∞ MIST (m)

PRECIPITATION

capital letter	signifies	Heavy	e.g. R
small letter	"	Moderate	e.g. r
suffix o	"	Slight	e.g. ro
prefix i	"	Intermittent	e.g. ir
doubling letter	"	Continuous	e.g. rr

	Light Intermittent	Light Continuous	Moderate Intermittent	Moderate Continuous	Heavy Intermittent	Heavy Continuous
DRIZZLE (d)	,	, ,	, ,	, , ,	, , ,	, , , ,
RAIN (r)	•	• •	• •	• • •	• • • •	• • • •
SNOW (s)	*	* *	* *	* * *	* * *	* * * *

SLEET

DEW (w)
HOAR FROST (x) no symbol used

SHOWERS (p) Slight Heavy Violent Snow Hail (h)

THUNDERSTORMS (tl) Slight or moderate Heavy With Rain/Snow With Hail

PAST WEATHER

Fair Variable sky Overcast Fog Drizzle Rain

Snow or sleet Showers Thunderstorms

FRONTS

	Printed Charts	Working charts
Warm Front (surface)		Continuous red line
Warm Front (above ground)		Broken red line
Cold Front (surface)		Continuous blue line
Cold front (above ground)		Broken blue line
Occlusion (surface)		Continuous purple line
Occlusion (above ground)		Broken purple line
Stationary Front (surface)		Alternate red and blue, continuous
Stationary Front (above ground)		Alternate red and blue, broken
Warm Occlusion (surface)		Continuous red line behind continuous purple
Cold Occlusion (surface)		Continuous blue line behind continuous purple
Lines of Frontogenesis		Corrugated red or blue line
Frontolysis	Short lines across Front	Short lines across Front, of same colour.

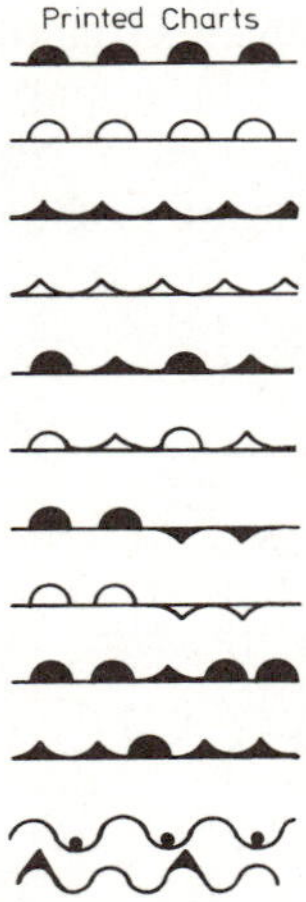

Note: symbols are placed on that side of line towards which Front is moving

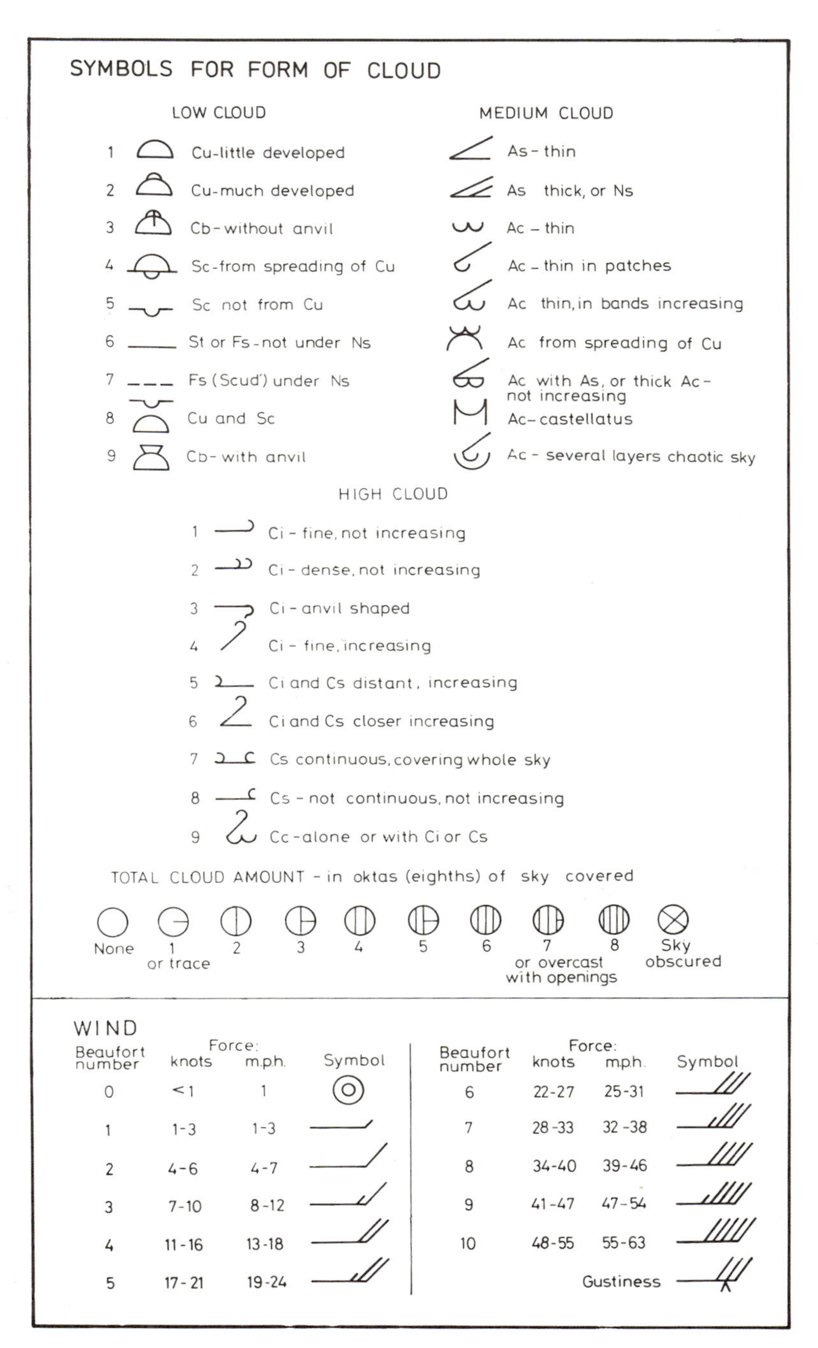

Fig. 9.3 (a) and (b). Symbols used on weather maps

the observing station is represented by a circle and the different elements are grouped around this circle. The symbols, letters, and figures used for various weather elements on synoptic charts are shown in Fig. 9.3. After all the available weather data for all stations have been plotted in the manner shown in Fig. 9.2, the analyst then proceeds to analysing the charts. The first step is to identify and mark special states of the atmosphere or of specific atmospheric processes according to conventional meteorological models (e.g. fronts).

Classical frontal analysis is based on the ideal cyclone model. Using this model it is believed that:

1 a front could be recognized from very few observations;
2 the configuration of front can be related to the life cycle of depression; and
3 weather condition can be classified according to air mass types.

Frontal analysis has, nowadays, lost some of its value primarily because of improvement in weather observations. It is now possible to have a picture of rain belts and other weather associated with fronts where the observation network is dense without recourse to any model front. Second, individual fronts often differ in characteristics from the model or ideal front and forecasters are more interested in these differences. Third, it is still very difficult to define a front exactly and clouds, precipitation, and several linear weather systems have been incorrectly labelled as fronts. Eight main types of front have been recognized. However, these are mere descriptions of the state at any given time of the general polar front between tropical and polar air masses. The names given to the fronts are even closely related to the evolution of the model depression. They are as follows: warm, cold, occlusion, back-bent occlusion, secondary cold, quasi-stationary, warm, and cold front waves. Temperature, dew point, and wind field are the three most useful parameters in identifying fronts on the weather chart although all observations are of help.

Once the fronts have been identified and drawn in, the weather analyst then considers pressure observations. Pressure values are reduced to mean sea level and isopleths drawn. Isobars are usually drawn at 4 mb or 5 mb intervals, but whatever the intervals the 1000 mb isobar is always included. The pressure field is analysed for two purposes. The first is to identify pressure systems which are often associated with particular types of weather. The second is to construct a picture of the wind field from a study of the pressure distribution using the geostrophic balance concept (see Fig. 5.1). This can only be done, however, in the middle latitudes. In the low latitudes, particularly around the equator, the coriolis force is very small tending towards zero as the equator is approached. There is, therefore, little or no geostrophic balance and the wind field can only be analysed directly from surface and upper air observations.

Upper air charts and diagrams are required to give the necessary third dimension to our picture of the atmosphere. The plotting model is similar to that used for surface charts and the major objective of upper air chart analysis is to draw isopleths to describe a meteorological field (Atkinson, 1968). Formerly, upper air charts were drawn for certain heights, namely 1 km, 3 km, 6 km.

More recently, upper air charts are drawn for given pressure (isobaric) surfaces. Such charts are known as contour charts. The World Meteorological Organization recommends that contour charts should be drawn for the following standard isobaric surfaces: 1000, 850, 700, 500, 400, 300, 200, 150, 100, 70, 50, 30, 20, and 10 mb. National meteorological services are enjoined to provide contour charts for at least four of the following standard isobaric surfaces: 850, 700, 500, 300, and 200 mb. This is to give a reasonably good three-dimensional picture of pressure distribution.

Contour charts have some advantages that recommend them for use among weather analysts and forecasters. One geostrophic wind scale can be used for all levels since the relationship between the geostrophic wind speed and the contour pattern is independent of air density. The same geostrophic scale can be applied to another useful type of upper air charts known as the *thickness chart*. A thickness chart is the chart on which values of the thickness of a given layer over an area are plotted. The usual thickness chart is for the layer 1000 to 500 mb. The difference between the heights of two pressure levels at a given place is known as the thickness of the layer of the atmosphere between those two levels. The thickness of a layer increases with the mean temperature of that layer. Thus, whereas the contour charts show us the circulation at the level for which they are drawn, thickness charts indirectly give us the temperature distribution throughout any defined layer since the thickness of any layer is directly proportional to the mean temperature of the air in the layer. The variations of temperature, wind, and humidity in the vertical can, however, be more fully investigated by analysing radiosonde observations. Two types of charts used in this exercise are the *hodograph* and the *tephigram*. These will now be described.

The hodograph

Although upper winds in the vertical can be displayed by means of a table or can even be plotted on surface synoptic charts, a wind hodograph gives a more complete and useful display of upper winds in the vertical. Apart from the actual winds at each level, information can be obtained on *vertical wind shear* and *thermal winds* between different levels. Vertical wind shear is the change of wind speed, direction, or both with height. Thermal winds are a measure of the wind shear, being the vector difference between winds at two levels (Wickham, 1970). The hodograph (see Fig. 9.4) consists of concentric circles at specified distances (representing wind speeds) from a central point. Winds at different pressure levels are drawn as vectors from the origin. Since the vectors are lines of appropriate length and direction, they give indication of the nature of change of wind direction and speed with height. Using the thermal wind concept (Fig. 9.5) we can make deductions concerning the distribution of warm and cold air in the atmosphere, and hence the changing thermal structure of the atmosphere.

The thermal wind concept illustrated in Fig. 9.5 is based on the assumption of geostrophic flow. If we know the magnitude and direction of wind at two

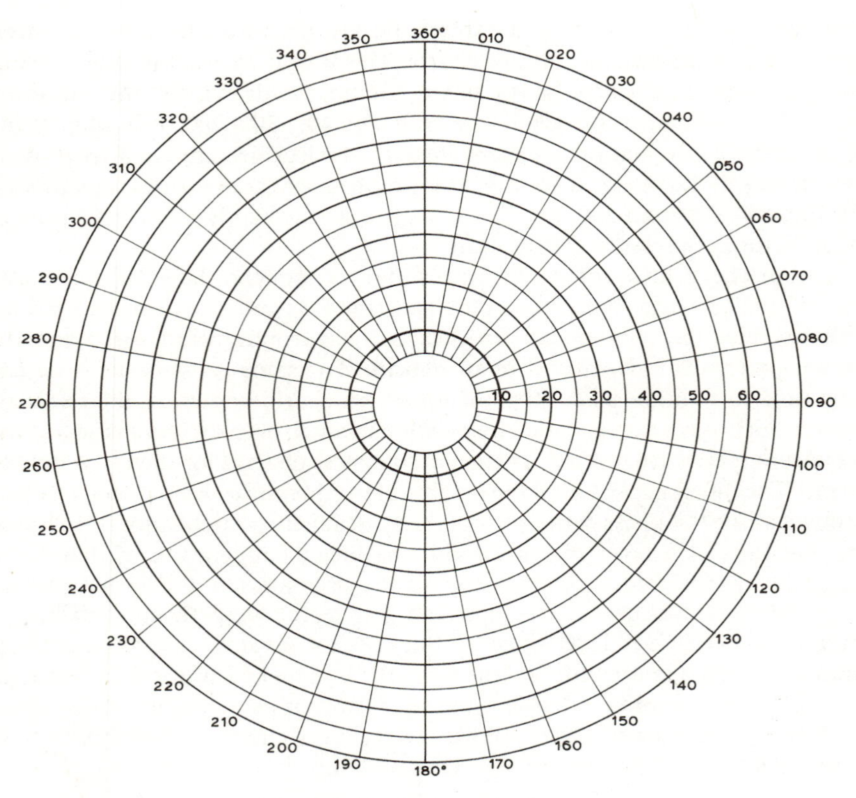

Fig. 9.4. The hodograph chart

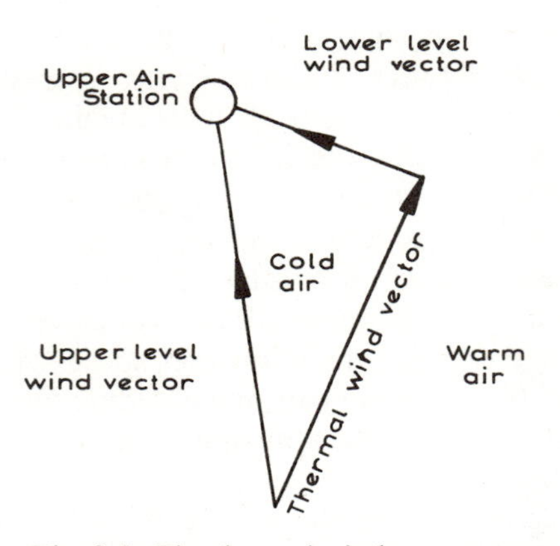

Fig. 9.5. The thermal wind concept

levels, by joining the end of the upper wind vector to that of the lower wind vector we would obtain the magnitude and direction of the thermal wind as shown in Fig. 9.5. Thermal wind in the northern hemisphere has cold air to the left and warm air to the right. The reverse obtains in the southern hemisphere. The location of the thermal wind thus enables us to establish the location of masses of relatively warm and cold air in the atmosphere and hence the stability or otherwise of the atmosphere.

The tephigram

The tephigram or aerological diagram is a thermodynamic diagram showing graphical representation of observations of pressure, temperature, and humidity made in a vertical sounding of the atmosphere. The diagram is a network of lines representing changes with height of five weather parameters: temperature, pressure, humidity, dry bulb potential temperature and wet bulb potential temperature (see Fig. 9.6). When dry bulb and dew-point temperatures from a radiosonde ascent are plotted on this diagram, the resultant curve together

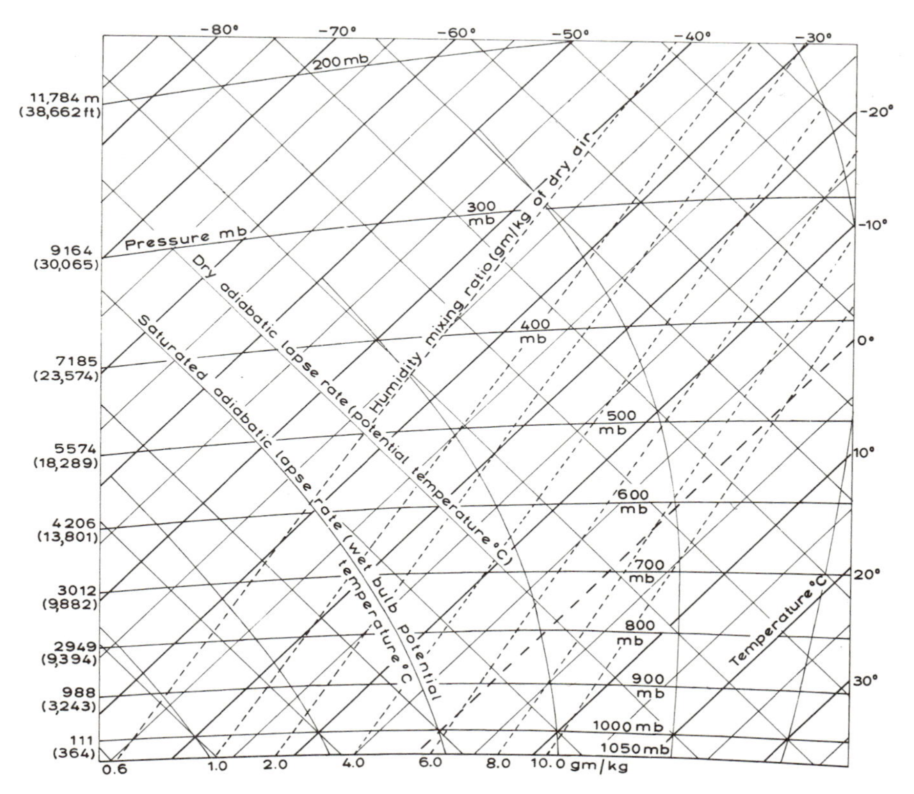

Fig. 9.6. A tephigram chart

with the path curve of ascending air parcel provides helpful information on the following:

1 identification of fronts whose presence is suggested by surface chart analysis;
2 air mass recognition is facilitated by the nature of the vertical distribution of wet bulb potential temperature which is little affected by evaporation or pressure changes;
3 analysis of the stability of the atmosphere;
4 estimation of potential depths of clouds.

As pointed out in Chapter 7, the stability or otherwise of a parcel of air depends on the relationship between the adiabatic and the environmental lapse rates. The dry adiabatic lapse rate is represented on the tephigram by straight lines of constant dry bulb potential temperature, sloping from bottom right to top left while the saturated adiabatic lapse rate is represented by lines of constant wet bulb potential temperature curving from bottom right to top left. If the environmental lapse rate is greater than the dry adiabatic lapse rate the rising parcel of air will be buoyant as it will always be warmer than the surrounding air. Conversely, if the environmental lapse rate is less than the dray adiabatic lapse rate the parcel of air will find itself moving into a warmer environment and will consequently lose its buoyancy and return to its original position.

The tephigram is a rather complicated graph with five sets of coordinates for five parameters as follows:

1 isotherms (lines of equal temperature) are the parallel lines running from bottom left to top right;
2 dry adiabats are the parallel lines from bottom right to top left;
3 isobars (lines of equal pressure) are the slightly curved nearly horizontal lines;
4 saturated adiabats are the curved lines sloping up from right to left;
5 saturated mixing ratio lines are those at slight angle to the isotherms.

The plotting of radiosonde records on the tephigram gives the environmental lapse rate curve. To determine stability condition, we have to plot the path curve for a rising parcel of air. This is indicated first by the dry adiabats and later by the moist adiabats when condensation has started taking place in the rising air. Once the environmental temperature curve and the path curve have been drawn in, stability is usually determined as follows:

1 the air is *stable* where the environmental temperature curve lines to the *right* of the path curve;
2 the air is *unstable* where the environmental temperature curve lies to the *left* of the path curve;
3 if the environmental and path curves are *identical* we have a *neutral* air (see Fig. 9.7).

The lifting condensation level can be determined using a tephigram. This is the level at which an air parcel becomes saturated if forced to rise. The point

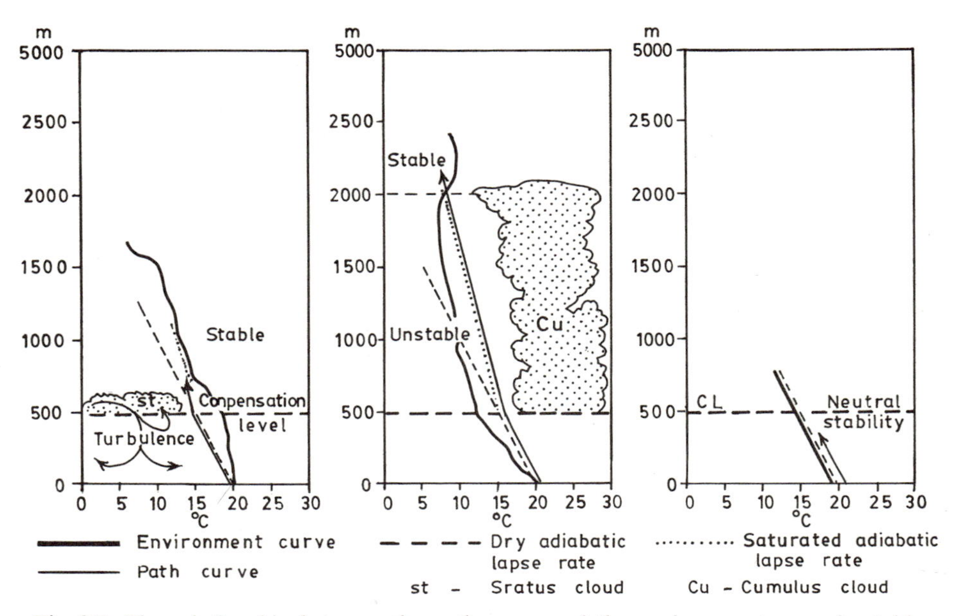

Fig. 9.7. The relationship between the path curve and the environment curve in stable, unstable, and neutral atmosphere (after Chandler, 1972)

at which a line along a dry adiabat through the surface air temperature value intersects a line along a saturation mixing ratio line through the dew-point temperature or a line along the saturated adiabat through the surface wet bulb temperature is the condensation level. While the lifting condensation level represents the lower limit of cloud development the level at which instability ceases in a rising parcel of air represents the general upper limit of cloud development. At this level, the environmental temperature curve, formerly on the left side of the path curve, intersects the latter. Other charts used in vertical analysis of weather include vertical cross-sections and frontal contour charts.

Vertical cross-sections are used to clarify the vertical structure of such weather phenomena as fronts and wind maxima (see Atkinson, 1968). The upper air stations are plotted on the horizontal axis and the information for the vertical axis is obtained from the tephigram. The frontal contour chart is a contour map of a frontal surface in which pressure is used as the vertical ordinate. The intersection of the upper surface of the frontal zone with the various standard pressure levels is traced and put on a map. Frontal zones have to be identified first using tephigrams and hodographs. Vertical cross-sections are rarely used on a routine basis as they take some time to prepare. On the other hand, frontal contour charts are much used in Canada by forecasters for day-to-day weather analysis.

Techniques of weather analysis involve a great deal of drudgery and subjectivity. The mechanics of weather chart production are now automated thanks to computers. Computers are now being used at an increasing number of national meteorological headquarters for processing observational data, plotting the data, and drawing isopleths. Weather maps which would take at least 30 minutes to construct by hand, are now being produced with the aid of computers in less than half a minute.

Principles of weather forecasting

Forecasting has undoubtedly been the greatest stimulus to the study of weather and climate. To many people, weather forecasting is even synonymous with meteorology. Man's desire to forecast or modify weather is age old. This is not surprising considering the fact that man and his activities are greatly influenced by the vagaries of weather. The term 'forecast' was first applied in meteorology by Admiral Fitzroy and it signifies a statement of anticipated meteorological conditions for a specified place, area, or route during a specified period of time. Weather forecasts are often classified into the following three types on the basis of the period covered by the forecasts:

1 short-period forecasts for part or whole of a 24-hour period with a further outlook for the following 24 hours;
2 medium-range forecasts are made for period of two to five days ahead;
3 long-range forecasts are made for period longer than five days ahead (e.g. month or season).

Although somewhat different techniques are used for the various types of forecasts, the basic problem of forecasting underlines all three types. This problem is twofold. First, we must know the state of the atmosphere at any given moment. Second, we must know the physical laws which govern the changes of that state. Two difficulties are encountered in our attempt to solve the above problem. One is that we do not have enough information to adequately characterize, in necessary detail, the existing state of the atmosphere. The other is that the mathematical equations expressing the physical laws are too complex for an exact solution unless some simplifications are made. This means that total success in forecasting can only be attained if and when these difficulties are completely resolved. Some of the details of the present state of the atmosphere are just unobservable while many equations expressing the physical laws have to be simplified for solution. Meteorology is, thus, far from being an exact science. Complete and accurate forecasts are impossible and forecasts will always have to be expressed in terms of probabilities. All we can do is to make the forecasts as accurate as possible. Obviously, the longer the period for which forecasts are made, the greater the chances of the forecasts going wrong.

Before discussing some of the methods of weather forecasting it is important to emphasize the fact there is a close relationship between weather analysis described earlier and weather forecasting. Both are usually done by the same person—the weather forecaster. The distinction between weather analysis and weather forecasting here is therefore somewhat artificial since the latter logically follows the former. Also, weather forecasting, like weather analysis, calls for skill and experience. The forecast is the result of the interplay of ideas and procedure in the forecaster's head, using all available evidence.

Methods of weather forecasting can be classified into three main types. These are:

1 synoptic methods,
2 statistical methods,
3 physical or numerical methods.

Synoptic forecasting entails the diagrammatical representation of weather systems through time and the extrapolation of developments of such systems into the future (Atkinson, 1968). The aim in synoptic forecasting is to produce from a synoptic chart of an existing situation, a similar chart portraying the circulation at a certain time in future. Such a chart of future surface circulation of the atmosphere is known as the *prebaratic or prognostic chart* while charts showing the future state of the upper atmosphere are known as *prontours*.

A prebaratic chart can be prepared by extrapolating recent changes into the future. The motion of existing weather systems can be extended into the future on the assumption that changes which have been observed to occur in the past will continue to occur in a similar way. Although this assumption may be reasonable over a period of up to 12 hours, for longer periods it will certainly not be valid. This is because the rate of motion of the systems may change. Besides, extrapolation of present trends cannot take account of new developments. Consequently, the technique of simple extrapolation can rarely be used with confidence in forecasts for 24 hours or more ahead. The synoptic charts for such forecasts are usually prepared with the aid of the computer. With computer charts available to the forecaster the necessary interpolations are made between existing conditions and those computed for 24 hours or so ahead.

The synoptic models—the ideal synoptic systems—are also used by forecasters in estimating the likely development of circulation patterns. Though there are many variations in these models, particularly those of frontal depressions and anticyclones, they are still useful tools in weather forecasting. Weather does not change in a completely random fashion. If it did, it would have been impossible to forecast weather. Rather there is a tendency for certain weather changes to follow each other in a fairly orderly succession. This characteristic of weather has given rise to the concept of a synoptic model. It must be borne in mind, however, that synoptic models are essentially descriptions of the more usual features of weather. The unusual features are excluded but if and when these occur the standard model breaks down.

Another technique of weather forecasting under the synoptic method uses *analogues*. Analogues are synoptic charts of actual past situations that are similar to current situations. Future developments may be forecast if such analogues are found and the evolution of past situations carefully studied. The problem with the use of analogues is that there is no perfect analogue. An infinite variety of synoptic patterns of atmospheric circulation is possible and does exist so that even reasonably good analogues are difficult to find. Besides, the comparison of synoptic patterns should really be three-dimensional. Records covering several decades are required in the search for analogues and so the exercise involves much labour. The analogue method can really only be used where the search for suitable analogues can be conducted with the aid of the electronic computer. The search for analogues forms much of the basis for most long-range monthly forecasts in many countries for want of a better technique.

Statistical methods are also used in long-range weather forecasts for a month or season. The Indian Meteorological Services, for example, use linear regression equations to forecast the coming of the monsoon and the amount of monsoon rains. Statistical methods are widely employed in long-range weather forecasts and in climate forecasting. The approaches available have been summarized by Lamb (1972) as follows:

1 understanding the relationship between the physical and especially the thermal condition of the earth's surface and the large-scale characteristics of the atmospheric circulation over a run of seasons or years;
2 study of the specific effects on atmospheric circulation and weather of anomalies of sea temperature and sea ice;
3 study of the effect of various kinds of disturbances on the sun on a variety of time scales on atmospheric circulation;
4 identification of the effects of volcanic and other dust and pollution veils on atmospheric circulation and weather and climate;
5 identification of the time scales, manner of operation as well as the physical origins of rhythmic or cyclic tendencies in weather and climate;
6 determination of statistical persistence tendencies in climatological values;
7 establishing statistically sequential tendencies (probable successions);
8 identification of analogous situations in the past and monitoring of how far similar development follows in present and past cases.

All the above approaches make extensive use of statistical methods, particularly of correlation and regression. Analogous cases for monthly forecasting and beyond are also selected by a variety of computerized scoring systems including correlation techniques although the final selection is made by forecasters after taking into account the physical controls believed to be working on the circulation. Finally, the success of forecasts is assessed after the event by a numerical scoring system or formulae. There are many of such formulae of which a simple example is as follows:

$$F_s = 100\left(\frac{R - E}{T - E}\right)\% \qquad (9.1)$$

Where F_s is the amount of skill shown in forecasts, R is the number of correct forecasts of an entity, E is the number expected to be correct if only persistence or climatological averages were used, and T is the total number of forecasts. This formula is, however, a bit severe on the forecaster since it gives him no credit for agreeing with the 'chance' forecast through his own independent physical reasoning. Some types of forecasting can only be made with the use of statistics (e.g. forecasts of probable frequency of heavy rainfalls in the future which engineers require for designing storm-drains). Statistical methods of forecasting are also known as the climatological methods. Long-range forecasts of weather always make comparisons with climatological normal conditions and it is usual to give forecasts of values of weather elements in terms of departures from normals.

Numerical forecasting

Numerical forecasting is based on the principle that the atmospheric circulation can be treated as a problem of fluid mechanics and thermodynamics. The atmosphere is regarded as a fluid of varying density that is unevenly heated and subject to the spin and frictional effects of the underlying earth (Atkinson, 1968). The problem numerical forecasting aims to tackle is twofold. The first is to quantitatively characterize the initial state of the atmosphere. The other is to apply the physical laws that control the changes of that state to predict future states. As mentioned earlier, there are difficulties in tackling this problem. There is a deficiency of data, particularly upper air data. The equations expressing the physical changes in the atmosphere are very difficult to solve exactly.

Numerical forecasting thus had to wait for the advent of electronic computers and the improvement of weather observations during World War II. Numerical forecasting was first suggested in 1912 as a possibility by V. Bjerknes but it was was not until 1922 that L. F. Richardson published the results of the first attempt at numerical forecasting. This attempt, though theoretically interesting, was a failure as a practical forecast. This was long before the advent of computers and well before upper air data from aircraft and radiosondes became more abundant in the 1940s.

Numerical forecasting is now done at the national meteorological headquarters of several developed countries. The following general steps are involved in numerical forecasting.

1 A grid is established over the area of interest. The grid usually covers a larger area than the forecast area to minimize errors arising from a grid boundary that artificially limits the horizontal extent of the atmosphere.
2 Values of pressure, temperature, etc., are then assigned to the grid intersection by statistical manipulation of observed data from surface and upper air station networks.
3 The most recent forecast of the present state of the atmosphere is fed into the model in terms of numbers on a grid. Forecast observations are usually given less weight than observed ones except on oceans.
4 Using the various equations describing physical laws that govern atmospheric motion (e.g., equations of motion, continuity, and state and the thermodynamic equation), future states of the atmosphere are computed and presented in map form.

Computed forecasts thus provide, at present, a framework of predicted contour lines and isobars. Weather details like cloud types, sunshine, rain, frost, fog, etc., still have to be supplied by the forecaster who will then fit them into the framework supplied by the computed forecast (Wickham, 1970). The new computational techniques and the traditional expertise of the forecaster are still both required. Also, the success of all forecasts, whether they are derived by computation or by non-computational methods, depends on the quantity and quality of the basic observational data utilized in preparing the forecasts.

Problems of weather forecasting in the tropics

Though weather in the tropics is less changeable than that in extratropical areas, the forecasting of weather in the tropics is beset with some peculiar problems. The observational networks in the low latitudes are far less dense than those in the middle latitudes and there is a dearth of information on the upper atmosphere. Although information from weather satellites is helping to fill the gaps created by poor networks of stations, there appears to be no alternative to drastic improvement in the basic surface and upper air stations networks. The range of instruments available at these stations must also be improved. For instance, very few stations have weather radar while radiosonde ascents are few and far between.

The problems of forecasting in the tropics are also less amenable to mathematical treatment than those in the temperate region. There is no simple approximate relation between wind and pressure distribution in the low latitudes. In fact, well defined pressure systems are usually absent in the tropics and the isobars bear hardly any relationship with the weather unlike in the temperate latitudes where forecasting is largely based on the association between isobaric forms and the types of weather that accompany them. Rather, in the low latitudes many of the manifestations of weather are instability phenomena incapable of precise prediction (Forsdyke, 1949). The geostrophic balance concept is inapplicable in the low latitudes because of the very low values of coriolis force. Isobars do not, therefore, give an indication of the speed and direction of winds in the upper atmosphere. These have to be directly observed or measured. Some basis other than isobars is therefore necessary for synoptic analysis in the tropics. This is provided by streamline analysis. A streamline is a line drawn parallel to the instantaneous direction of the wind vector at all points along it. A streamline map gives an instantaneous picture of the field of motion. Streamlines may be obtained directly from wind observations or may be computed from pressure distribution. Although streamlines only give an indication of wind direction, wind velocities may be estimated by means of the stream function which is related to streamlines in much the same way as the geostrophic wind is related to isobars. Using these estimated values of wind velocity or measured values, if available, lines of equal wind speed called *isotachs* are then drawn.

Also, in the tropics, covergence and divergence can only be estimated from the observed wind field unlike in the temperate region where they can be assumed as they occur in association with certain features of isobaric and isallobaric fields and with fronts. Surface synoptic charts must, therefore, be supplemented with upper air charts in the tropics. The usefulness of barometric tendencies for forecasting in the low latitudes is also limited. This is because values of barometric tendencies are rather low and are often not much greater than observational errors in barometric readings.

It is clear from the foregoing that some of the techniques used for weather forecasting in the temperate region are inapplicable in the tropics or must be greatly modified to be applicable there. In the tropics upper wind charts have to be constructed from observed data using radiosondes. Tephigrams are also not as useful in the tropics as in the middle latitudes. This is mainly because there are no well defined changes of air mass type in the tropics. Consequently, the markedly characteristic tephigrams associated with various air mass types in the temperate region are non-existent in the tropics. Rather, tephigrams tend to be characteristic of locality and season while day-to-day fluctuations are very small. Suitably located sferic or radar stations are required in the tropics for the detection of thunderstorms, so common there and which constitute the principal hazard to aviation in the low latitudes.

References

Atkinson, B. W. (1968). *The Weather Business*. Aldus Books, London.

Barrett, E. C. (1974). *Climatology From Satellites*. Methuen, London.

Chandler, T. J. (1972). *Modern meteorology and Climatology*. Thomas Nelson, London.

Forsdyke, A. G. (1949). Weather forecasting in tropical regions. *U. K. Meteorological Office Geophysical Memoirs No. 82*.

Lamb, H. H. (1972). Problems and practice in longer-range weather and climate forecasting. In Taylor, J. A. (ed.). *Weather Forecasting For Agriculture and Industry*. David and Charles, Edinburgh.

Wickham, P. G. (1970). *The Practice of Weather Forecasting*. H.M.S.O., London.

CHAPTER 10

Climatic Variations and Climatic Change

Introduction

The atmosphere is never static. Rather, it is in constant turmoil. Atmospheric characteristics change from place to place and over time at any given place on time scales ranging from microseconds to hundreds of years. There are important interactions within the atmosphere resulting from or causing such changes. These are appropriately called feedback mechanisms since there are no simple one-way cause and effect processes as effects often rebound to alter their causes. Thus the changes within the atmosphere may be internally induced within the earth–atmosphere system or externally induced by extraterrestrial factors.

It is important to make a distinction between *weather variations* and *climatic variations.* Weather is extremely variable particularly in the temperate region. But whether in the tropics or in the temperate region the existence of diurnal and seasonal weather changes cannot be denied. The weather changes collectively make up climate. There are evidences of fluctuations or variations in climate itself. When these fluctuations follow a trend we talk of *climatic trends.* The fluctuations may also be cyclical in nature to give what are called *climatic cycles.* Over a long period of time climatic fluctuations may be such that a shift in type of climate prevailing over a given area takes place. In that case, we talk of a change in climate or *climatic change.* The various terms used to describe variations in climate, namely climatic variability, climatic fluctuations, climatic trends, climatic cycles, and climatic change refer to some appropriate time scales and can only be validly used within such time scales.

One nomenclature which has been suggested for climatic change on various time scales is shown in Fig. 10.1. First, we have variability in climate that is too rapid to be regarded as climatic change. Such variability includes fluctuations in climate within a period of less than 30–35 years, a period usually used in calculating values of climatic normals. Second, there are secular or instrumental changes in climate occurring over a period of 100–150 years. Third, there are variations in climate during historical time dating back to a few thousand years. Finally, we have variations in climate on geological time scales running into millions of years.

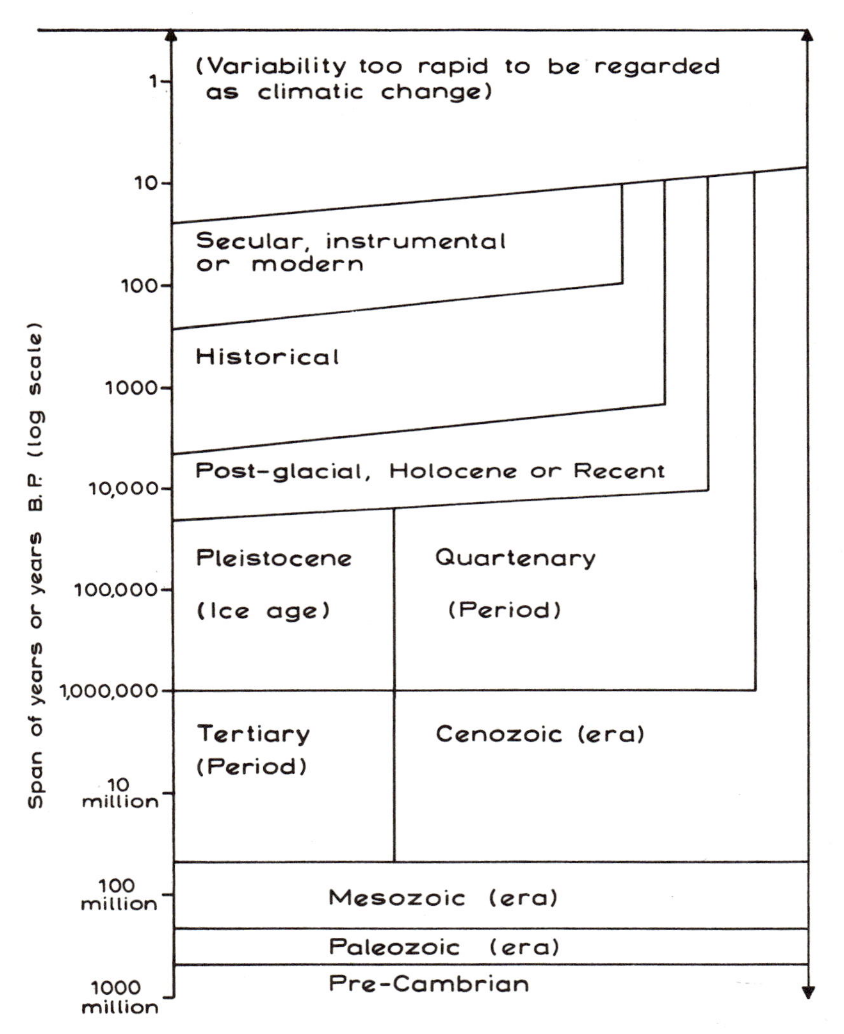

Fig. 10.1. Climatic change nomenclature

Indicators of past climates

Past climatic changes on the above different time scales are studied using different techniques and evidences. The discussion of past climates in this book is organized in two parts. The first deals with past climates during a geological period before recorded history. The second part deals with climate during recorded history. Our knowledge of the climate prevailing before recorded history comes from indirect sources of evidence in the earth's crust. Such evidences of past climates are many and varied. They can, however, be grouped into three broad categories, namely biological, lithogenetic and morphological.

Biological indicators of past climates include fossils (remains of ancient plants and animals preserved in sedimentary rocks), pollens, and tree rings. Lithogenetic indicators of past climates include varves, salt deposits and other sedimentation phenomena, weathering phenomena like laterites, glacial deposits like moraines and fossil soils, among others. Morphological indicators of past climates include inselbergs, river terraces, fossil dunes, and glacial features like corries and eskers and relict landforms.

Biological indicators of past climates

The main biological indicators of past climates are fossils, pollens, and tree rings. Fossils found in sedimentary deposits can be dated using standard paleontological techniques like carbon-14 dating. The fossils and the sedimentary deposits in which they are found are studied to determine the duration and spatial extent of temperature and moisture conditions that favoured the existence of such organisms (plants and animals) where they have been fossilized. There are, however, uncertainties involved in interpreting or using fossils as indicators of past climates. Fossilization is selective so that the fossils found may not be very representative of the floral or faunal assemblage that existed in the past in a given location. Also, the climatic requirements of a plant or an animal may change or vary in the course of time. It is quite possible that extinct species of plant/animal were adapted to a climate quite different from that in which their modern counterparts are found. Finally, fossils may be derived so that they indicate climatic conditions which belong to an earlier period (Schwarzbach, 1963).

Pollens are fine powdery substances discharged from the anther of flowers of plants. Pollens are blown by wind and may settle on land and water surfaces. If pollens settle on a water surface they sink to the bottom of the lake or swamp and are incorporated in a sedimentary layer. Pollens that settle on land will most likely decay. Pollens found in sedimentary rocks can be analysed and dated. The technique of pollen analysis is known as *palynology*. By comparing pollens found in sedimentary rocks with pollens of modern vegetation it is possible to infer the climates of the times when the pollens were preserved. Climatic changes are reflected in the pollen spectra of successive sedimentary layers. The two major limitations of pollens as indicators of past climates are:

1 that like fossils pollens are selectively preserved and may not be representative of the vegetation existing at the time;
2 pollens, like fossils, may also be derived so that they indicate climatic conditions which in fact belong to an earlier period.

Analysis of tree rings may be used to detect climatic changes in the last few centuries or so. This procedure is known as *dendrochronology*. In areas with regular seasonal changes of weather trees generally produce one growth ring per annum. The thickness of the growth ring varies with climatic conditions. The growth ring is very thick when climatic conditions are optimum for the growth of the trees while the growth ring is narrow if conditions are not favourable. Dendrochronology is useful in detecting climatic changes only during the very recent past because most trees grow to old age within the relatively short period of a few hundred years. Also, the thickness of growth rings is influenced not only by climatic conditions but also by the age of the trees which has nothing to do with climate.

Lithogenetic indicators of past climates

Lithogenetic indicators of past climates often used in studies of climatic change include varves, evaporites, weathering processes, particularly lateritization, and their products. Varves are distinctive banded layers of silt and sand deposited annually in lakes ponded near the margins of ice sheets. The coarser material which is also lighter in colour settles first during summer melting while the finer and darker material is deposited in winter. Each band of light and dark material represents one varve. By counting the number of varves the number of years involved in the formation of a varve deposit can be estimated. The relative thickness of each individual year's sedimentation can be used to infer variations in climatic conditions throughout the period of deposition.

Evaporites or salt deposits can only occur under dry warm conditions in which evaporation exceeds precipitation. The existence of these deposits in areas now humid and cool is an indication that the climate has been drier and warmer in the past. Laterites are also associated with warm humid or seasonally humid environments. Their occurrence therefore suggests that such conditions had prevailed in the past in the areas where they are now found.

Morphological indicators of past climates

Morphological evidences of past climates are many and varied. They include relict land forms (e.g. old beaches and sand dunes and glacial land forms like moraines and eskers) and river terraces. Relict land forms do not generally present problems of interpretation once they are recognized. Sand dunes are features of an arid environment and so suggest aridity. Similarly glacial features like moraines and eskers suggest the occurrence of glaciation in the past. However, old beaches may have come about for reasons other than climatic change. A fall in the sea level caused by epeirogenesis, for example, will cause the creation of new beaches as the old ones become located farther away from the coast. Similarly, river terraces can be created by climatic change involving variations in the discharge/load relationship of the river as well as by processes of river capture and tectonism which have nothing to do with a change in climate.

Evidences of climatic change must therefore be carefully studied and evaluated before they are accepted as such. Besides, various evidences which corroborate each other should normally be used in determining the pattern of changes in climate in the past. It is often misleading to rely on one evidence alone in determining the pattern of climate in the past. The study of past climate is rather like solving a jig-saw puzzle. It requires patience. All available evidences have to be considered together to arrive at reasonable decisions concerning the pattern of climate in the past. The various evidences of climatic change result from the effects of climate on animals, vegetation, and land form, among others. What then are the causes of changes in climate? It is to this important question that we now direct our attention.

Causes of climatic change

A change in climate implies a change in the general circulation of the atmosphere on which climate ultimately depends. Climate, however, involves not only the atmosphere but also the hydrosphere, the biosphere, the lithosphere, and the cryosphere. These are the five components that make up the climatic system. This system is also subject to extraterrestrial influences particularly that of the sun. Climate therefore depends on or is determined by two major factors. These are:

1. the nature of the five components making up the climatic system and the interactions between the various components;
2. the nature of the geophysical conditions outside the climatic system and the influence they exert on the climatic system.

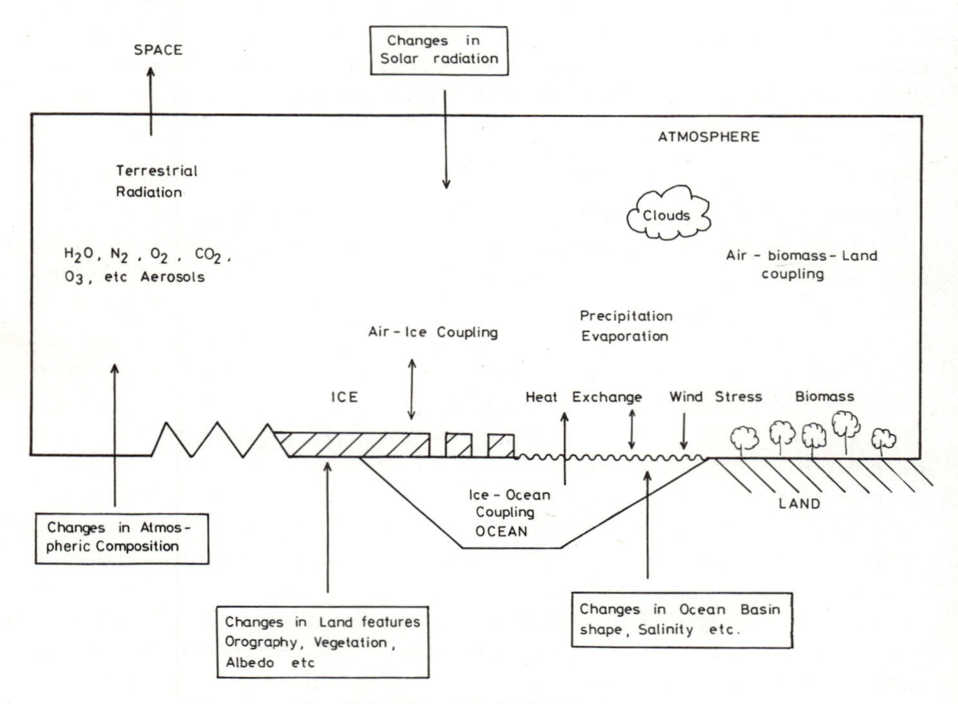

Fig. 10.2. The global climatic system

Table 10.1 Theories of causes of climatic change

Terrestrial causes
Polar wandering and continental drift
Changes in the earth's topography
Variations in atmospheric composition
Changes in the distribution of land and water surfaces.
Variations in snow and ice cover
Astronomical causes
Changes in the eccentricity of the earth's orbit
Changes in the precession of the equinoxes
Changes in the obliquity of the plane of the ecliptic
Extraterrestrial causes
Variations in solar radiation amount (solar output)
Variations in the absorption of solar radiation outside the earth's atmosphere

Fig. 10.2 is a schematic illustration of the climatic system with some examples of the physical processes responsible for climate and climatic change. The climatic state at any given period depends on three crucial factors. These are:

1 the amount of energy received by the climatic system from the sun;
2 the manner in which this energy is distributed and absorbed over the earth's surface; and
3 the nature of the interaction processes between the various components of the climatic system.

Theories of climatic change attempt to account for variations in the above three factors over time. Variations in climate, however, occur on different time scales and we may therefore require different theories to explain such variations. This is one reason why no single theory of climatic change has been found satisfactory in explaining all the variations that have been known to occur in world climate. Besides, it is believed that various factors operate to bring about a change in climate. Any theory of climatic change must therefore take a broader view of the mechanisms of change in climate. The various theories of climatic change that have been put forward by various workers over the years can be discussed under three broad categories, namely terrestrial, astronomical and extraterrestrial causes of climatic change (see Table 10.1).

Terrestrial causes of climatic change

Theories of climatic change under this group attempt to link changes in climate to variations in terrestrial conditions. Changes in the distribution of land and oceans would bring about a change in energy distribution and hence general atmospheric circulation and climate because of the well known differences in the thermal characteristics of land and water surfaces. Examples of such theories include those of polar wandering and continental drift. Shifts in the locations of continents and oceans would also mean that given areas would be located nearer or farther away from the poles or equator with concomitant changes in climate. There are also theories relating to changes in the topography of the continents and oceans particularly the former. Mountain building processes may be expected to influence climate in two ways. First, the topography is changed with concomitant changes in the influence exerted on air flow, insolation, and other weather elements like temperature and precipitation. Second, orogenesis may involve volcanicity which would provide aerosols and other pollutants that will affect the transparency of the atmosphere and consequently the amount of energy reaching or leaving the earth's surface. All these will have effects on the earth's energy balance and hence climate.

Several other terrestrial theories of climatic change are based on variations in atmospheric transparency. Apart from volcanic aerosols mentioned earlier, there have been variations in the following atmospheric constituents—carbon dioxide (CO_2), ozone (O_3), and water vapour. All these constituents play important roles in the energy balance of the earth. Variations in their concentrations in the atmosphere may therefore be expected to influence global energy balance and consequently the general circulation of the atmosphere on which climate depends.

Astronomical causes of climatic change

Astronomical theories of climatic change are based on changes in the earth's geometry. The following are the major ones: changes in the eccentricity of the earth's orbit, in the precession of the equinoxes, and in the obliquity of the plane of ecliptic. Fluctuations of the eccentricity of the earth's orbit cause variations in the receipt of solar energy by the earth. The distance of the sun from the centre of elliptical orbit controls the distance of the earth from the sun at different times of the year as well as the duration of the four seasons. The smaller the eccentricity of the elliptical orbit the smaller will be the differences in the length of the seasons and the greater the eccentricity the greater will be the variations between the seasons. At the perihelion when the earth is nearest the sun, solar energy receipt is 6% more than at the aphelion when the earth is farthest away from the sun. The eccentricity of the earth's orbit fluctuates with a periodicity of about 92,000 years. This means that in about 50,000 years the earth in its orbit will be nearest to the sun in July and not in January as at present. Summers in the northern hemisphere may therefore become warmer and winters colder during the next 50,000 years.

The precession of the equinoxes also varies with a periodicity of about 22,000 years. The term refers to the regular change in the time the earth is at a given distance from the sun. At present the equinoxes occur on 21st March and 23rd September while the solstices occur on 21st June (summer solstice) and 21st December (winter solstice). The displacement of the four seasonal points will result in the migration of the seasons along the orbit. This displacement is believed to be caused by the gravitational attraction between the sun, the moon, and the earth. Fluctuations in the precession of the equinoxes will cause a shift in the seasons.

There are also variations in the obliquity of the plane of ecliptic with a periodicity of about 41,000 years. The obliquity is at present about $23\frac{1}{2}°$ but has varied in the past periodically from $21\frac{1}{2}°$ to $24\frac{1}{2}°$. The seasons result from this fact that the earth is inclined at this angle to its orbit round the sun. Thus, a decrease in the obliquity of the ecliptic would decrease the differences between seasons but increase the distinction of climatic zones. On the other hand, an increase in the angle would cause marked seasonal differences but geographical zones would be less distinct or even disappear (Gates, 1972).

Extraterrestrial causes of climatic change

Theories of extraterrestrial causes of climatic change postulate changes in the amount of solar energy reaching the earth either because of changes in the solar output or changes in the amount of solar radiation absorbed outside the earth's atmosphere. There are short term, medium term, and long term cyclical fluctuations in the amount of solar output. Sunspots cause fluctuations of output with cycles of 11 years, 22 years, 44 years, etc. Solar flares also cause short term fluctuations in the nature and amount of solar radiation. Such fluctuations are known to occur in the solar spectrum especially in the ultraviolet range and in the X-rays (cosmic radiation) which increase during solar flares. Tidal oscillations on the sun raised by the planets in their orbits also cause fluctuations in solar output. Finally, variations in intersteller dust particles cause variations in the amount of solar energy reaching the top of the earth's atmosphere. Of all these influences, the sunspots have been most intensively studied. Increases in sunspots have been associated with cooler and wetter conditions while decreases are associated with warm and drier conditions.

World climate during geological periods

It is now well established that climate has varied during the earth's history although the causes of these variations are still not fully understood as indicated above. Although the earth is estimated to be about three to four billion years old, the study of past climates (palaeoclimatology) covers only some 500–600 million years. This is because of the nature of evidence of past climates found only rarely in pre-Cambrian rocks. As mentioned earlier palaeoclimatology depends on fossils and some non-biological indicators of past climates while its dating techniques are borrowed from palaeontology—a branch of geology concerned with the study of fossils.

A summary of the history of climate of the world since the pre-Cambrian is presented in Fig. 10.3. Our knowledge of world climate improves as we consider more recent periods. For instance more is known about climatic variations during the Pleistocene than during the Tertiary period. The greater part of the last billion years is believed to have enjoyed warm ice-free conditions. This

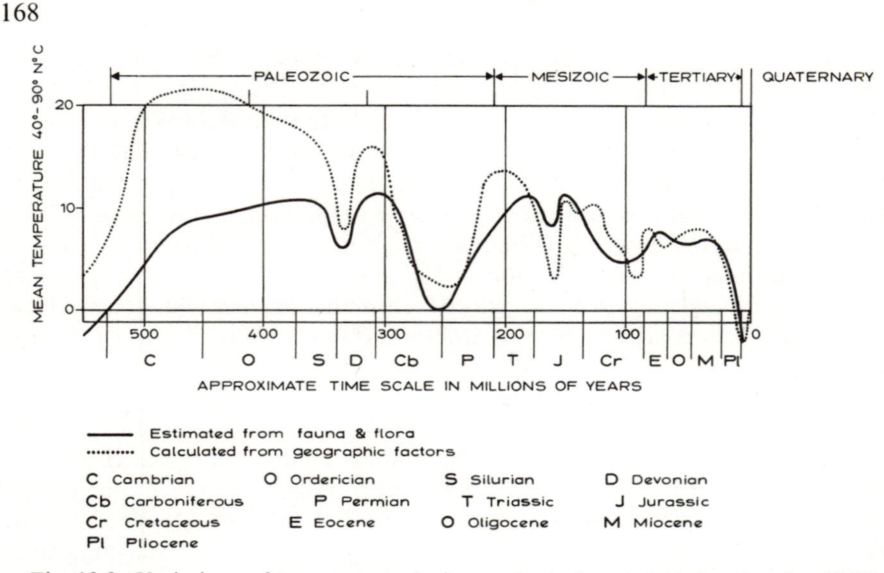

Fig. 10.3. Variations of temperature during geological periods (after Brooks, 1949)

warm climate was interrupted by two glacial ages before the Pleistocene epoch of the last million years. About 300–350 million years ago ice sheets covered present day South America, Africa, India, Australia, and the Antarctica. These land masses were joined together then in a supercontinent called 'Gondwanaland'. Earlier, between about 600–800 million years ago extensive glaciation occurred in Greenland, Africa, Australia, and Aisa. Fig. 10.3 shows a 150 million year period of high temperatures between 500 and 350 million years ago. Since then the general trend of temperature has been downward though there have been some notable large variations. The lowest temperatures in the northern hemisphere since the Cambrian prevailed during the Pleistocene glaciation of about a million years ago.

A fairly more detailed summary of the variations in climate over the earth from the pre-Cambrian period to present time is shown in Table 10.2. Three glaciation periods have occurred in the last 600 million years or so. There was glaciation in the pre-Cambrian period, during Permian and more recently during the Pleistocene. Fig. 10.4 shows the graph of air temperature over the earth in the last 850,000 years. It indicates that the earth's climate has been cold most of the last million years, swinging through a series of glacial and interglacial episodes during which ice sheets alternately advanced and retreated as the temperatures dropped and rose (Roberts and Landsford, 1979). The last period of extensive continental glaciation during the Pleistocene occurred between 22,000 and 14,000 years ago. The ice sheets then began to retreat. By 8500 years ago the ice sheets in Europe had retreated to their present positions while the North American ice sheets withdrew to their present positions about 7000 years ago.

The period from 7000 to 5000 years ago was warmer than the present time

Table 10.2 Climates of the various geological periods (after Brooks, 1949)

Era	Period	Age by radio activity in million years	Climate
Quaternary	Recent (Holocene)	1	Glaciation in temperate latitudes
	Pleistocene		
Tertiary	Pliocene	13	Cool
	Miocene	30	Moderate
	Oligocene	60	Moderate to warm
	Eocene	. .	Moderate becoming warm
Mesozoic	Cretaceous	110	Moderate
	Jurassic	155	Warm and equable
	Triassic	190	Warm and equable
Palaeozoic	Permian	210–240	Glacial at first, becoming moderate
	Carboniferous	260–300	Warm at first, becoming glacial
	Devonian	310–340	Moderate becoming warm
	Silurian	340	Warm
	Ordovician	400	Moderate to warm
	Cambrian	510	Cold, becoming warm
Pre-Cambrian		560	Glacial

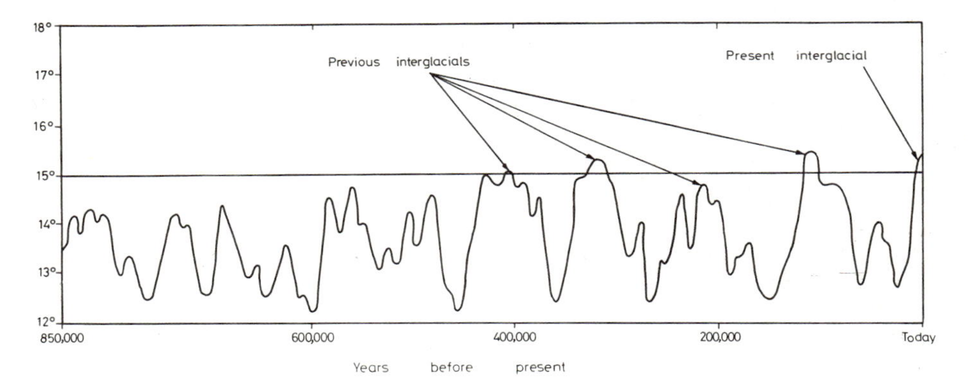

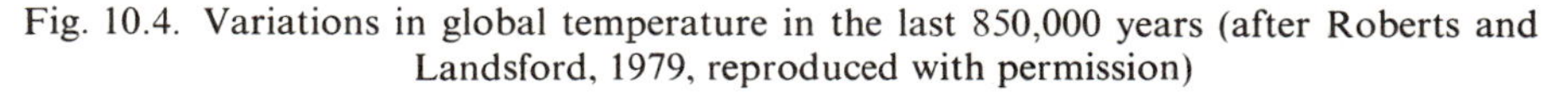
Fig. 10.4. Variations in global temperature in the last 850,000 years (after Roberts and Landsford, 1979, reproduced with permission)

but the last 5000 years have been marked by generally declining temperatures with exceptionally cold intervals about 2800 and 350 years ago. The latter cold interval between 1550 and 1850 has been called the 'little ice age'. During this period European glaciers grew and vineyards disappeared from England.

It is believed by many climatologists that we are at present passing through an interglacial period the length of which nobody really knows.

World climate during recorded history

Our knowledge and understanding of climatic variations during recorded history is by far much better than during the geological period. Evidences are obtained from a variety of sources including archaeology and anthropology, documentary evidence and lately instrumental records. During the entire period of written and inferred human history, the above sources serve to double check data from varve analysis, tree ring analysis, and other geochronological analysis concerning changes in climate.

Archaeological excavations have provided indications of what type of climate prevailed in the past over many areas. For instance, rock drawings of savanna animals have been used to show that the northern and southern peripheries of the Sahara desert were 100–250 km towards the interior about 7000 years ago. The distribution of artefacts also indicates that Neolithic man and his stock could move across the Sahara relatively easily at this time. One is therefore bound to conclude that no real desert existed then in parts of the present day Sahara desert. Ancient well sites and water works have also provided evidence that populations once lived in areas where life is now virtually insupportable. The oral history of famines and large-scale migration also indicates deterioration of climate in the past. These evidences have to be carefully considered, however, since famine or migration could be induced by war or pestilence that bears no relationship to deteriorating climatic conditions.

The pattern of climatic change in the tropics, particularly in Africa, over the last 2000 years has been outlined by Grove (1968, 1972). The following appears to be the sequence of climatic variations in Africa in the last 2000 years. About 20,000 to 15,000 years ago, the climate was drier than at present over large areas of tropical Africa, particularly in lands now semi-arid or subhumid. Rainfall was about a third of the present amount. It was also colder than now, temperatures being 4–6 °C lower then than at present. Evaporation losses were about two-thirds of the present values. The post-glacial rise in the sea level resulted in the formation of estuaries, barrier beaches, and enclosing lagoons.

About 12,000 to 7000 years before the present, the climate over most parts of Africa was warm and wetter than now. Temperatures were, however, lower than at present though higher than what obtained 20,000 to 15,000 years before the present. During this time, the southern Sahara probably enjoyed Mediterranean type of climate which persisted in the Chad area until 5000 years ago (Grove, 1972). The Sahara desert was much smaller in extent then than at present. As earlier mentioned, rock drawings of savanna animals in the present day Sahara desert indicate that the northern and southern boundaries of the Sahara were 100–250 km towards the interior some 7000 years ago. Lake Chad was then much bigger than now. Mega-Chad's level was 320–330 metres compared with Lake Chad's presented level of about 282 metres (Grove, 1972).

This pluvial period was succeeded by another dry period lasting a thousand years. This gave way in several areas to one or more phases when the climate was more humid than at present. The changes of climate over the last 2000 years have been minor, leaving little evidence on the landscape. This has made the study of climatic variations over the last 2000 years rather difficult until we reach the period of instrumental records in the late nineteenth century. Records of variations in lake or river levels for earlier times exist for some areas, notably the Nile Basin and East Africa.

In the case of Europe, however, instrumental meteorological records date back for about three centuries and in North America for about two centuries. In other continents including Africa they are confined to the last 100–150 years and in many cases, this century. It must be borne in mind, however, that the accuracy of the very early instruments is not always above reproach. This applies in particular to rainfall records because of the various ways the early rain gauges were exposed. Some of the early types of rain gauges did not make adequate provision against the re-evaporation of fallen rain.

A chronological arrangement of the variations in climate over Europe in the last 8000 years or so is shown in Table 10.3. A similar chronological list of variation in climate in northern Africa in the last 2500 years up to the fourteenth

Table 10.3 Fluctuations of climate in Europe (mainly after Brooks, 1949)

BC	
5400	Moist and warm
5000	Drier and cooler
4500	Moist, rather warm
4000–3000	Becoming cooler and drier
2200	Very dry especially in Central Europe
2000	Rainy period
1200–1000	Dry and warm
850	Somewhat moister and cooler
700–500	Dry and warm
500	Sudden, increase of rainfall, much cooler
AD	
0	Climate similar to present
100	Drier and warmer
180–350	Wetter
600–700	Dry and warm
800–1200	Little ice, rainfall heavy in Central Europe
1200–1300	Great storminess, mild winters, probably rainy
1500–1600	Dry
1600	Beginning of general advance of glaciers
1677–1750	Dry period, generally mild winters
1850–1950	Rise of winter temperature, general recession of glaciers
1950–Present	Slight cooling trend resulting in decrease in the length of the growing season

Table 10.4 Variations of climate in northern Africa (after Brooks, 1949)

Date	Kharga Oasis	Nile	Other localities
500 BC	Prosperous	Floods high	
AD 100	Excessive well boring	Floods good	Alexandria rainier in summer
AD 200	Temporary decline of prosperity	Floods poorer	
AD 400	Improvement	Floods good	Mandingan Empire AD 320–680
AD 700	Great decline	Minimum level AD 700–1000	Mandingan Empire broke up
AD 1150	Almost depopulated	Rise of low level stage around AD 1100	
AD 1225	More prosperous	Level low	Traffic in now waterless Eastern Desert
AD 1300	Still further improvement	Level low	Prosperous Sudanese states

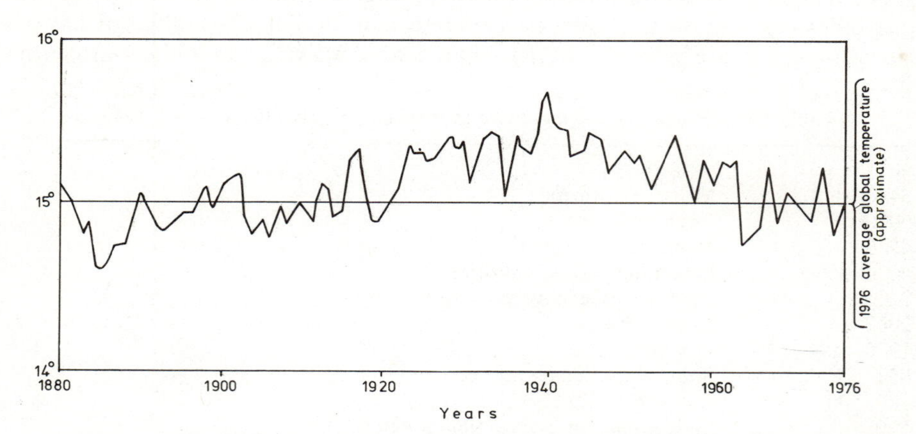

Fig. 10.5. Temperature variations in the northern hemisphere in recent times (after Roberts and Landsford, 1979, reproduced with permission)

century is shown in Table 10.4. Similar chronological arrangements of variations in climate over Asia and America during the period of recorded history have been attempted by various workers (see Brooks, 1949).

The pattern of temperature variation in the northern hemisphere since instrumental records started is shown in Fig. 10.5. There was a global warming trend beginning in the 1880s and terminating in the 1940s when cooling set in. The cooling trend appears to have been reversed since the 1960s. The trend of temperature is important particularly in extratropical areas where temperatures determine the length of the growing season. Precipitation has also varied since instrumental records began as studies from various regions indicate. In West Africa, for example, major droughts have occurred at least three times since the beginning of this century. There were droughts in 1913–14, 1943–6 and recently 1972–4. The 1913–14 drought affected most parts of Africa. Similarly the 1972–4 drought in the West Africa sahel extended to Ethiopia and parts of East Africa.

Since instrumental records of weather elements began, wide variations in rainfall have similarly been experienced in North America, Europe, and Asia. Severe droughts were experienced in the United States in 1893–4, in the 1930s and in 1975–6. Europe has also experienced several droughts in the past the most recent being the drought of 1975–6. Southeast Asia has also suffered from rainfall variations giving periods of droughts and floods. The failure of the seasonal monsoon rains is the main cause of drought in this part of the world. Floods, on the other hand, are caused by cyclones or bursts of monsoon rains. These extremes of moisture supply are well exemplified in the Indian subcontinent where flood and drought disasters occur rather frequently.

It is quite apparent from the above that the study of climate of the past requires the skill of the geologist, botanist, and climatologist, among others, and requires patience and dedication. Various sources of evidence have to be carefully considered and reconciled. A supposedly climatic trend may in fact be part of a climatic cycle. A shift in climate may even be more difficult to determine. While temperature or rainfall decreases over time in a given area the trend at another place may be opposite. Local climatic trends, cycles, or even changes may be out of tune with regional or continental overall pattern of climatic fluctuations. Techniques of analysing the evidences of past climates are being improved and are now very sophisticated. Variations in climate in recent times are better documented thanks to the improvement in the network of weather stations although there is still room for further improvement. The low latitudes, the polar areas, the deserts, and the oceans are still poorly served by weather stations.

References

Brooks, C. E. (1949). *Climate Through the Ages* (2nd edn). Dover Publications, New York.

Gates, E. S. (1972). *Meteorology and Climatology for the Sixth Form and Beyond* (4th edn). Harrap, London.

Grove, A. T. (1968). The last 20,000 years in the tropics. In Harvey, A. M. (ed.). *Geomorphology in a Tropical Environment*. British Geomorphological Research Group Occasional Paper No. 5.

Grove, A. T. (1972). Climate change in Africa in the last 20,000 years. In *Les Problèmes de Développement du Sahara*. Septrional Vol. 2 Algiers.

Roberts, W. O. and Landsford, H. (1979). *The Climate Mandate*. W. H. Freeman, San Francisco.

Schwarzbach, M. (1963). *Climates of the Past: An Introduction to Palaeoclimatology*. Van Nostrand, London.

W. M. O. (1979). *Proceedings of the World Climate Conference* W. M. O., Geneva.

CHAPTER 11

Classification of Climate and Regional Climates

Introduction

The variations of climate over time were examined in the previous chapter. In this chapter, emphasis will be on the pattern of spatial variations in climate. Although no two locations on the earth's surface have identical climates, it is possible to define areas within which climate is broadly uniform from place to place. Such a region is usually known as a *climatic region*. Regional climatology is the branch of climatology concerned with the identification, mapping, and description of climatic regions over the earth or part thereof.

The climate over a given location is the synthesis of all the climatic elements in a somewhat unique combination determined by the interplay of climatic controls and processes. There is therefore a wide variety of climates or climatic types prevailing over the earth's surface. To facilitate the mapping of climatic regions, these numerous climates must be classified using suitable criteria. Climatic classification therefore emerges from the need to synthesize and group similar climatic elements into climatic types from which climatic regions are mapped.

Purpose and problems of climatic classification

The overriding purpose of any classification system is to obtain an efficient arrangement of information in a simplified and generalized form. The purpose of climatic classification is to provide an efficient framework for organizing climatic data and learning about the complex variations in world climate. Through climatic classification the details and complexities of monthly or seasonal climatic statistics are compressed into simpler forms which are more easily understood. In short, we classify to simplify, clarify, and understand the world's complex climatic patterns.

Climatic classification though desirable is a difficult exercise. Certain problems faced in climatic classification are common to all types of classification whether of soils or vegetation. We have to acknowledge the fact that all classifications are artificial to the extent that we impose order (boundary) on a complexity or continuum. Consequently, many classifications are subjective. In fact classification is more a product of human ingenuity than a natural phenomenon. In climatic classification we face other problems which emanate from the inadequacy of available climatic data both in terms of coverage of the earth and in terms of duration and reliability. Climate is also dynamic, not static, so it fluctuates and varies over time. This of course implies that our climatic boundaries will also fluctuate. Finally, climate is a multivariate phenomenon consisting of various climatic elements. There is therefore the problem of identifying the crucial climatic parameters that constitute distinctive climatic types. The climatic elements most frequently used to characterize the climate over a given area are temperature and rainfall. Often only the average values of these elements are employed in the classification exercise. The need to consider other climatic elements cannot be overemphasized. Also average values of climatic elements must be considered alongside departures from such averages as extremes may be more significant limiting values. To overcome the problems created by the multivariate nature of climate, some classification schemes have taken the natural vegetation as an index of the climatic conditions prevailing over an area. Several non-climatic factors, however, exert control over the character of the vegetation in a given area. Such factors include topography, soil type, and the effects of human activities like farming and lumbering.

Approaches to climatic classification

There are today several schemes of classifying climates. The value of any climatic classification must be judged by its success in achieving the objective for which it has been designed. There are two fundamental approaches to classification of climates: the genetic approach and the generic or empiric approach. In the first approach, classification is based on the climatic controls. These are the factors that determine or cause the different climates. Examples include air circulation patterns, net radiation, and moisture fluxes. In the second approach, classification is based on the observed climatic elements themselves or their effects on other phenomena, usually vegetation or man. Those climatic classifications based on the influence of climate on man are really biometeorological in nature and are therefore specialized classifications. Examples include those based on human physiological comfort, building types to maintain optimum physiological comfort or clothing requirements. (See Chapters 4 and 13). Because the controls of climate are far more difficult to measure than the climatic elements there is a paucity of suitable data for most parts of the earth. Most climatic classification schemes have therefore adopted the empirical approach for which data are more available. Thus, in a recent survey of 169 climatic classification schemes only 21 could be regarded as genetic in approach, the remaining 148 schemes being empirical in approach (Terjung and Louie, 1972). The majority of the genetic classifications schemes deal with broad dynamic–synoptic systems with only two concerned with the energy budget of the earth's surface.

Genetic classification schemes

Four examples of climatic classification schemes adopting the genetic approach are briefly reviewed here. Two of these are based on broad dynamic–synoptic systems while the remaining two are based on the energy budget. A genetic classification of climates suggested by H. Flohn in 1950 recognizes seven climatic types on the basis of global wind belts and precipitation characteristics (see Table 11.1). Temperature does not appear explicitly in the classification. Another simple though extremely effective genetic classification of world climates was proposed in 1969 by Strahler (see Table 11.2). World climates are divided into three major divisions–the low latitude climates, the middle latitude climates and the high latitude climates. These three divisions are then subdivided into 14 climatic regions to which is added high land climates having altitude as the dominant control.

In 1956, Budyko proposed a simple but highly generalized classification of climate based on the energy balance. The classification is based on values of the radiational index of dryness (Id) defined by the equation of the form

$$Id = \frac{Rn}{Lr} \tag{11.1}$$

where Rn is the amount of radiation available for evaporation from a wet surface assumed to have an albedo of 0.18, L is the latent heat of evaporation

Table 11.1 Flohn's genetic classification of climates (Flohn, 1950)

The two main criteria used in the classification scheme are:

1. the global wind belts, and
2. precipitation characteristics.

Temperature does not appear explicitly in the classification scheme.

Climatic type	Precipitation characteristics
I Equatorial westerly zone	Constantly wet
II Tropical zone, winter trades	Summer rainfall
III Subtropical dry zone (trades or subtropical high pressure)	Dry conditions prevail throughout the year
IV Subtropical winter rain zone (Mediterranean type)	Winter rainfall
V Extratropical westerly zone	Precipitation throughout the year
VI Subpolar zone	Limited precipitation throughout the year
VIa Boreal continental subtype	Summer rainfall limited; winter snowfall
VII High polar zone	Meagre precipitation; summer rainfall, early winter snowfall

Table 11.2 Strahler's genetic classification of climates (Strahler, 1969)

The criteria used in this classification scheme are:

1 the character of the dominant air masses, and
2 precipitation characteristics.

I Low latitude climates—controlled by equatorial and tropical air masses:
(a) Wet equatorial
(b) Trade wind littoral
(c) Tropical desert and steppe
(d) West coast desert
(e) Tropical wet–dry

II Middle latitude climates—controlled by tropical and polar air masses:
(a) Humid subtropical
(b) Marine west coast
(c) Mediterranean
(d) Middle latitude desert and steppe
(e) Humid continental

III High latitude climates—controlled by polar and arctic air masses:
(a) Continental subarctic
(b) Marine subarctic
(c) Tundra
(d) Ice cap
(e) Highland climates—major highland areas of the world where altitude is the dominant control of climate

and r is the mean annual precipitation. The value of the radiational index of dryness (Id) is less than unity in humid areas and greater than unity in dry areas. Using this index the following five major climatic types were recognized by Budyko (1956):

Climatic type	Radiational index of dryness (Id)
I Desert	> 3.0
II Semi-desert	2.0 –3.0
III Steppe	1.0 –2.0
IV Forest	0.33–1.0
V Tundra	< 0.33

The above classification scheme gives only a generalized picture of the world climates owing to the nature of the index and the fact that very few stations in the world have reliable data on the net radiative flux.

A more rigorous and detailed genetic classification of world climates based on the fluxes of energy and moisture was recently proposed by Terjung and Louie (1972). The following simplified version of the energy budget equation was used in the exercise

$$Rn + F_{\downarrow} = LE_{\uparrow} + H_{\uparrow} + F_{\uparrow} \tag{11.2}$$

where Rn is the net radiation, $F_{\downarrow}$ is horizontal sensible heat importation, $LE_{\uparrow}$ is the latent heat of vaporization, $H_{\uparrow}$ is sensible heat flux removing energy from the interface (convection) and $F_{\uparrow}$ is the horizontal sensible heat exportation. The seasonal march of the parameters of the above equation covering twelve months was determined for 1058 stations widely located over the globe. After careful analysis of the various graphs, six major climatic groups and 62 climatic types were identified (see Table 11.3).
The six major climatic groups are as follows:

A—Climates: macrotropical climates
B—Climates: subtropical climates
C—Climates: middle-latitude-continental climates
D—Climates: mesotropical climates

Table 11.3 Energy input–output climates of the world according to Terjung and Louie (1972)

Energy input determines the major climatic groups and subdivisions are made on the basis of energy output.
The major climatic groups are as follows:

A Climates (macrotropical climates)—macro energy input, micro range
B Climates (subtropical climates)—macro energy input and medium range
C Climates (middle-latitude-continental climates)—large input, large range
D Climates (mesotropical climates)—medium input, very low range
E Climates (marine-cyclonic climates)—medium input, medium output
G Climates (polar climates)—minimal input and range

Below is the matrix of the total 62 input–output climates obtained.

Input climates	Output climates		Transitional oceanic				
	Wet	Dry					
A	A	A	BE	CC	DA	DD	DG
	D	B	CB	CC	GD		
			EE				
B	A	B	BE		DA	DD	DG
	B	C	CC		GA	GD	GG
	D	C	EB	EE			
			GB				
C	B	C	EC	EE	DA		
	C	E			EB		
	E				GA	GB	GG
	G						
D	A	D	AE		DG		
	D		CE		GD	GG	
			EE				
E	GE		BF				
G			CF				
			DE				
			GD	GE	GG		
G	G		GE	GG			

E—Climates: marine-cyclonic climates
G—Climates: polar climates

Although a global classification of climates based on the input of energy and the subsequent disposal of that energy at the interface is good in theory, a major shortcoming of such a classification scheme derives from the general lack of accurate and reliable data on the energy balance components for most parts of the world.

Empirical classification schemes

Of the numerous climatic classification schemes that adopt the empirical approach only three will be briefly reviewed here. These are those of Koppen developed between 1900 and 1936, Thornthwaite (1948) and Miller (1965). Koppen's climatic classification scheme is relatively simple and very popular. For more than four decades now, most textbooks on regional geography and

Table 11.4 Koppen's climatic classification scheme

The most used system of climatic classification is that of W. Koppen (1846–1940), either in its original form or with modifications. Koppen himself modified and revised his classification first put forward in 1900. Koppen's scheme basically relates climate to vegetation but numerical criteria are used to define climatic types in terms of climatic elements. Koppen's first classification scheme in 1900 was based on vegetation zones shown on the world vegetation map of Alphonse de Candolle, a French plant physiologist. This scheme was revised in 1918 with greater attention given to temperature, rainfall, and their seasonal characteristics.

Koppen's classification scheme has five major climatic types recogniged on the basis of temperature and designated by capital letters as follows:

A Tropical rainy climates
B Dry climates
C Warm temperate rainy climates
D Cool snow-forest climates
E Polar climates

To these is added a group of undifferentiated highland climates represented by the symbol H. Each of A, B, C, D, and E climates is further subdivided using additional temperature and rainfall characteristics as set out below.

A TROPICAL RAINY CLIMATES
- Af Tropical rainforest climate
- Aw Savanna climate
- Am Tropical monsoon climate

B DRY CLIMATES
- BSh Hot steepe climate
- BSk Cool steppe climate
- BWh Hot desert climate
- BWk Cool desert climate

C WARM TEMPERATE RAINY CLIMATES
- Cfa Moist in all seasons, hot summer
- Cfb Moist in all seasons, warm summer

Table 11.4 (*Contd.*)

Cfc Moist in all seasons, cool, short summer
Cwa Summer rain, hot summer
Cwb Summer rain, warm summer
Csa Winter rain, hot summer
Csb Winter rain, warm summer

D COOL SNOW-FOREST CLIMATES
Dfa Moist in all seasons, hot summer
Dfb Moist in all seasons, cool summer
Dfc Moist in all seasons, cool summer
Dfd Moist in all seasons, severe winter
Dwa Summer rain, hot summer
Dwb Summer rain, warm summer
Dwc Summer rain, cool summer
Dwd Summer rain, severe winter

E POLAR CLIMATES
ET Tundra
EF Perpetual snow and ice

The major categories (i.e. A, B, C, D, E) are based mainly on temperature criteria as follows:

A Coldest month has a mean temperature greater than 18 °C. The 18 °C winter isotherm is critical for the survival of certain tropical plants. The annual rainfall is greater than the annual evapotranspiration.
B Mean annual potential evapotranspiration is greater than the mean annual precipitation. There is no water surplus, hence no permanent streams originate here.
C Coldest month has a mean temperature between − 3° C and 18 °C. Warmest month has a mean temperature greater than 10 °C. The 10 °C summer isotherm correlates with the poleward limit of tree growth and the − 3 °C isotherm indicates the equatorward limit of permafrost.
D Coldest month has a mean temperature below − 3 °C and warmest month has a mean temperature greater than 10 °C.
E Warmest month has a mean temperature less than 10 °C. ET warmest month has a mean temperature between 0 °C and 10 °C. EF warmest month has a mean temperature less than 0 °C.

Subdivisions of each major category are made with reference to:

1 the seasonal distribution of precipitation
f = no dry season, wet all the year (A, C, D)
m = monsoonal, with a short dry season and heavy rain during the rest of the year (A)
w = summer rain (A, C, D)
S = summer dry season (B)
W = winter season (B)
2 additional temperature characteristics
a = hot summer, warmest month has a mean temperature greater than 22 °C
b = warm summer, warmest month has a mean temperature less than 22 °C
c = short cool summer, less than 4 months have mean temperature greater than 10 °C
d = very cold winter, coldest month has a mean temperature less than − 38 °C

In the arid regions (BW and BS), the following subscripts are used:

h = hot, mean annual temperature greater than 18 °C
k = cool, mean annual temperature less than 18 °C

climatology have adopted Koppen's method of climatic classification or a modification of it. This is not to say that the scheme is perfect as will be shown later. In Koppen's scheme there are five major climatic groups recognized mainly on the basis of temperature characteristics. These five groups are further subdivided on the basis of the seasonal distribution of precipitation and additional temperature characteristics to give a total of 24 climatic types (see Table 11.4).

Despite its quantitative approach, objectivity, and invaluable merits as a teaching device several criticisms have been made of the scheme. The scheme has been criticized for lacking a subhumid category, for being empirical rather than genetic, for lack of justification for the use of some of the numerical criteria and even for the use of rigid boundary criteria at all in the light of the dearth of climatic observations over most parts of the world. Koppen's climatic classification has particularly been critized by Thornthwaite who described it as unsystematic, being based on a patchwork of unrelated rules and definitions. Koppen's climatic regions are in essence vegetation regions climatically defined. Miller's scheme (Miller, 1965) is rather similar to Koppen's scheme. There are seven major groups recognized using temperature criteria and these are again subdivided on the basis of rainfall characteristics to give 19 climatic types (Table 11.5). Many of the criticisms of Koppen's scheme are also applicable to Miller's scheme because of the obvious similarity between the two schemes.

In 1948, Thornthwaite proposed a rational classification of climates based on the concept of potential evapotranspiration and the water balance (see Chapter 7) and a moisture index derived from purely climatic data. Thornthwaite's approach to climatic classification has been found useful in diverse fields such as ecology, agriculture, and water resources development. Thornthwaite's classification scheme has, however, not escaped criticism. Values of potential evapotranspiration on which the moisture index is based are not readily available for many areas. Thornthwaite's empirical formula for estimating values of this parameter has failed to produce satisfactory estimates over many parts of the world particularly the arid and semi-arid areas. Thornthwaite's moisture index has been shown to be inconsistent, structurally weak, and theoretically unsound (see Ayoade, 1972). The ratio of actual evapotranspiration to the potential has been shown to be a better index of delimiting moisture regions at least in Nigeria. Thornthwaite's method of classification is rather unwieldy, and according to Thornthwaite himself lacking in mathematical elegance and without a nomogram would have been difficult to use. Let us now consider the classification scheme which is set out in some detail in Table 11.6.

Thornthwaite's classification is based on two main climatic indices—the moisture index and the annual potential evapotranspiration. The moisture index (Im) is given as follows

$$Im = \frac{100S - 100D}{PE} \tag{11.3}$$

where s is the annual water surplus, D is the annual water deficit, and PE is the

Table 11.5 Miller's climatic classification scheme

Miller's climatic classification scheme is based on temperature and rainfall criteria. Two considerations that underline the scheme are:

1 the seasonal distribution of rainfall, particularly the length of the dry season, if any;
2 the seasonal distribution of temperature and particularly the length of the cold season, if any.

Because of the obvious importance of the first factor in low latitudes, it is used to subdivide the hot climates. Similarly, the latter factor is used to subdivide the temperate and cold climates.

Seven major categories of climates are recognized and each of these except mountain climates is further subdivided. The major climatic categories are represented by diagnostic letters as follows:

A Hot climates
B Warm temperate climates
C Cool temperate climates
D Cold temperate climates
E Arctic climates
F Desert climates
G Mountain climates

The various subdivisions are set out below along with the criteria employed:

A HOT CLIMATES (mean annual temperature greater than 21.1 °C: no month has mean temperature less than 18 °C)
- 1 Equatorial double maxima rain
- 1m Equatorial monsoon type
- 2 Tropical marine: no marked dry season
- 2m Tropical marine: monsoon type
- 3 Tropical continental: summer rain
- 3m Tropical continental: monsoon type

B WARM TEMPERATE CLIMATES (no cold season: no month has mean temperature less than 6.1 °C)
- 1 Western margin (Mediterranean): winter rain
- 2 Eastern margin: uniform rain
- 2m Eastern margin (monsoon type): marked summer maximum rain

C COOL TEMPERATE CLIMATES (cold season of 1–5 months having mean temperature less than 6.1 °C)
- 1 Marine: uniform rain or winter maximum
- 2 Continental: summer maximum of rain
- 2m Continental: (monsoon type): strong summer maximum rain

D COLD CLIMATES (long cold season 6 months or more having mean temperature less than 6.1 °C)
- 1 Marine: uniform rain or winter maximum
- 2 Continental: summer maximum of rain
- 2m Continental (monsoon type): strong summer maximum rain

E ARCTIC CLIMATES (no warm season – 3 months or less have mean temperature above 6.1 °C)
- Ice climates (always cold; no month has a mean temperature above 6.1 °C)

F DESERT CLIMATES (mean annual precipitation is less than one-fifth of the mean annual temperature)
- 1 Hot deserts: no cold season—no month has mean temperature less than 6.1 °C
- 2 Mid-latitude deserts: one or more months have mean temperature less than 6.1 °C

G MOUNTAIN CLIMATES

Table 11.6 Thornthwaite's rational classification of climates (Thornthwaite, 1948)

In Thornthwaite's classification, potential evapotranspiration (*PE*), precipitation (*P*) and the relationship between these variables provide the bases of four climatic criteria used in the classification. The four criteria are each represented by an index value and boundaries are set quantitatively. The four criteria are:

1 moisture adequacy represented by values of the moisture index;
2 thermal efficiency represented by the values of potential evapotranspiration;
3 seasonal distribution of moisture adequacy; and
4 summer concentration of thermal efficiency.

Climatic moisture type		Moisture index
A	Perhumid	100 and above
B_4	Humid	80–100
B_3	Humid	60–80
B_2	Humid	40–60
B_1	Humid	20–40
C_2	Moist subhumid	0–20
C_1	Dry subhumid	– 33.3–0
D	Semi-arid	– 66.7 to – 33.3
E	Arid	– 100 to – 66.7

Thermal Efficiency and its summer concentration

Type	Thermal Efficiency	*PE* (cm)	Summer concentration Type	Concentration (%)
A′	Megathermal	114 and above	a′	Below 48.0
B'_4	Mesothermal	99.7–114.0	b'_4	48.0–51.9
B'_3	Mesothermal	85.5– 99.7	b'_3	51.9–56.3
B'_2	Mesothermal	71.2– 85.5	b_2'	56.3–61.6
B′	Mesothermal	57.0– 71.2	b′1	61.6–68.0
C'_2	Microthermal	42.7– 57.0	C_2'	68.0–76.3
C'_1	Microthermal	28.5– 42.7	C′1	76.3–88.0
D′	Tundra	14.2– 28.5	d′	above 88.0
E′	Frost	below 14.2		

Seasonal moisture adequacy is determined for the moist climates by the values of the aridity index given as

$$\frac{D}{PE} \times 100$$

where *D* is water deficit and *PE* is potential evapotranspiration. The seasonal moisture adequacy for the dry climates is determined by the values of the humidity given as

$$\frac{S}{PE} \times 100$$

where *S* is the water surplus and *PE* is potential evapotranspiration.

Table 11.6 (*Contd.*)
Seasonal Moisture Adequacy

Moist climates (A, B, C_2)		Aridity index
r	little or no water deficit	0–10
s	moderate summer deficit	10–20
w	moderate winter deficit	10–20
s_2	large summer deficit	above 20
w_2	large winter deficit	above 20
Dry climates (C, D, E)		**Humidity index**
d	little or no water surplus	0–16.7
s	moderate winter surplus	16.7–33.3
w	moderate summer surplus	16.7–33.3
s_2	large winter surplus	above 33.3
w_2	large summer surplus	above 33.3

annual potential evapotranspiration. Values of water surplus and water deficit are obtained by method of climatic water budgetting (see Thornthwaite and Mather, 1955) while values of potential evapotranspiration can be estimated using a temperature-based empirical formula (Thornthwaite, 1948). Moisture provinces are delimited on the basis of the computed values of the moisture index (Table 11.6). Thermal provinces are similarly defined by the use of the annual potential evapotranspiration which is regarded as an index of available energy.

Using these two indices and additional criteria 120 climatic types were hypothesized by Thornthwaite (1948) of which only 32 could be shown on the world map. These subdivisions are too many compared to Koppen's 24. To simplify the scheme and for certain purposes (e.g. agriculture and hydrology) only the moisture index is often used. World maps of the distribution of climatic types whatever the classification scheme used indicate that climatic types tend to be zonal in distribution (see Figs. 11.1 and 11.2). This indicates the dominant role of radiation (and latitude which primarily controls it) in climate.

In concluding this brief survey of approaches to climatic classification, mention should be made of recent attempts to classify climates using multivariate statistical techniques notably factor analysis and principal components analysis. One major problem of climatic classification stems from the fact that climate, as a synthesis of weather, is a multivariate phenomenon requiring the use of many climatic variables to characterize it. The use of multivariate statistical techniques enables us to consider simultaneously as many climatic variable as we wish as criteria for classification. The numerous variables are collapsed into a few components or factors which account for most of the variance in the original data matrix. Because of the great deal of computations involved, the widespread use of multivariate techniques had to await the advent of the electronic computer in the 1950s.

One of the earliest empirical classification of climates using multivariate

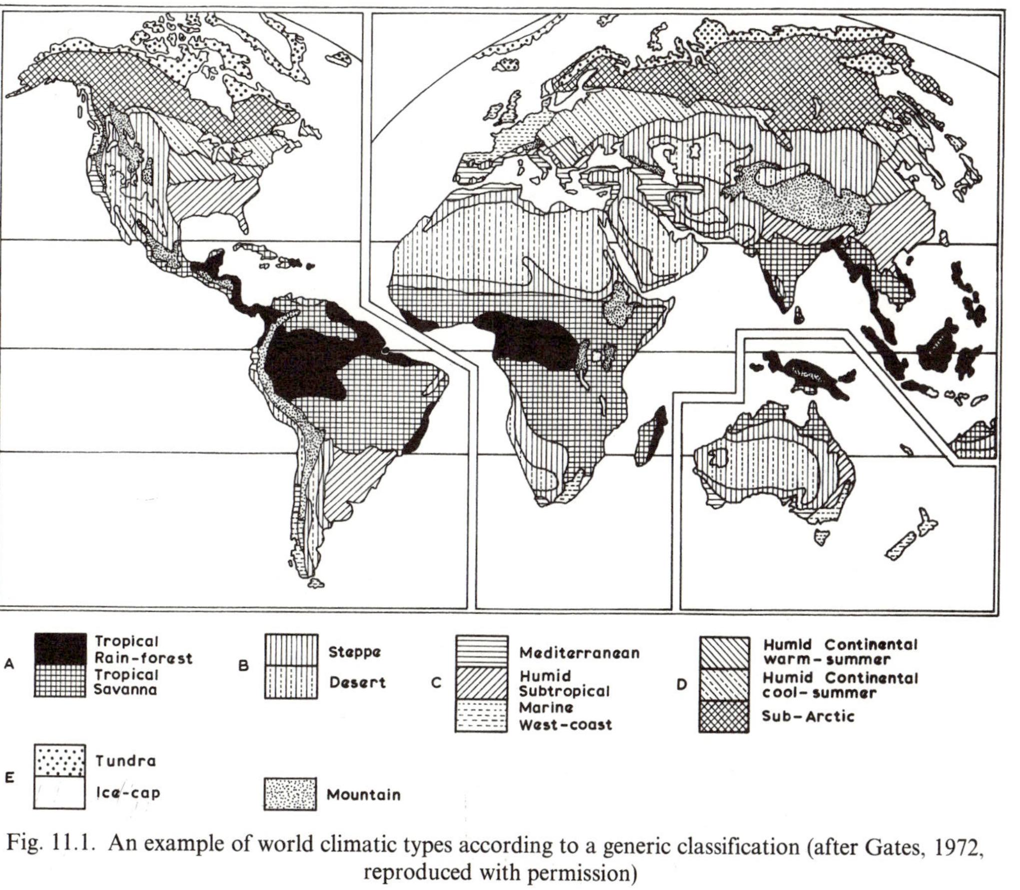

Fig. 11.1. An example of world climatic types according to a generic classification (after Gates, 1972, reproduced with permission)

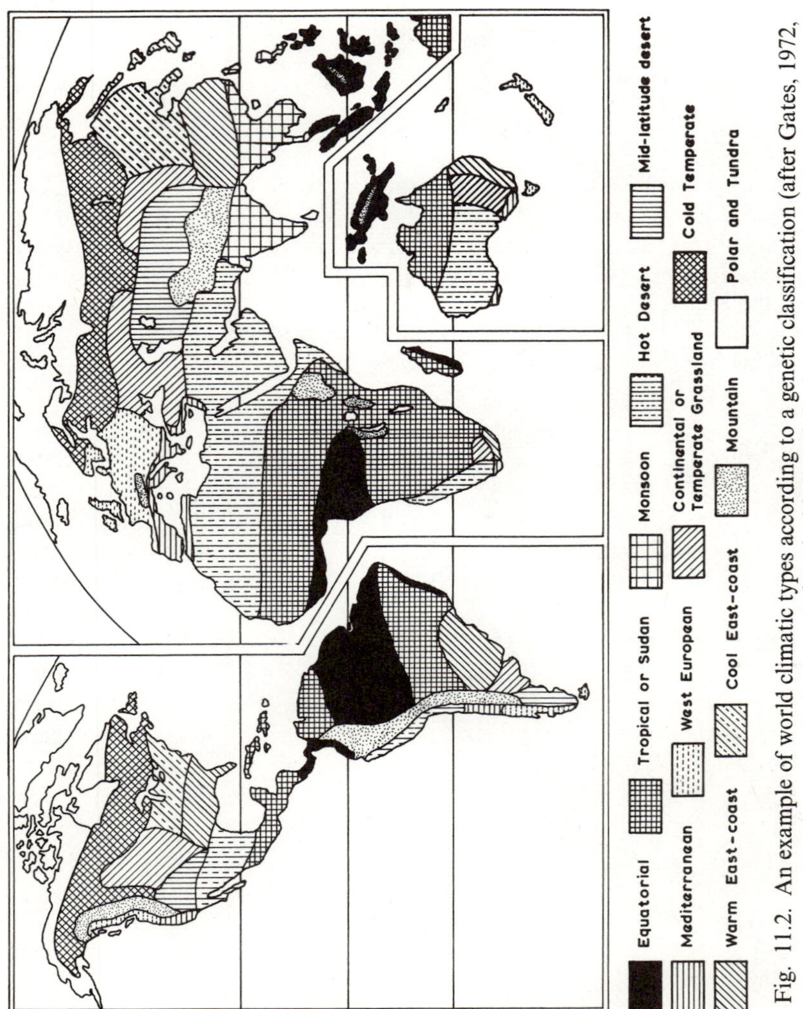

Fig. 11.2. An example of world climatic types according to a genetic classification (after Gates, 1972, reproduced with permission)

statistical techniques was that by Steiner (1965). Sixteen weather variables for 67 weather stations in conterminous states in the United States were collapsed into four components which account for about 87% of the variance in the original matrix. The sixteen variables represent various characteristics of four major weather elements namely temperature, precipitation, sunshine, and humidity. Apart from mapping component scores and interpreting them in meteorological terms, regionalization is done by cluster analysis using distance measures calculated from component scores of weather stations as indices of similarity of the climate at these stations. Since Steiner's work, similar attempts have been made to classify the climates of Australia, Europe, South Africa, and Nigeria. The use of multivariate statistical techniques in climatic or other classifications, however, has its own problems and limitations (see Johnston, 1969 and Ayoade, 1977). Different groupings are obtained using different grouping algorithms. The weather variables used in the classification exercise also influence the climatic types that are eventually obtained. The number of groups obtained is also largely subjectively determined by the individual after taking into account the purpose of the classification among other things.

Approaches to regional climatology

Regional climatology is essentially descriptive in approach. The climate of a given region is described with the aid of graphs of seasonal variations in the values of the climatic elements, usually temperature and precipitation. Little or no attempt is made to explain the observed climate. But recently increasing emphasis has been given to the explanation of the observed climatic patterns. This has involved relating observed weather and climate to the pattern of the prevailing atmosphere circulation—the so-called synoptic approach.

The spatial unit adopted in regional climatology has also varied from one author to another. Some discuss regional climatology within the framework of a climatic classification scheme. Climatic regions are delimited and their climates described in turn with the aid of characteristic graphs of seasonal variations in temperature and rainfall. Others discuss regional climatology on the basis of the continent rather than a world-wide climatic region or belt. The climate of a given continent is first broadly described and explained with reference to the atmospheric circulation systems. This is followed by discussion of the regional variations in climate within the continent.

In this book, we are primarily concerned with the principles of climatology and the bases for climatic classification rather than regional climatology as described above. There are excellent texts on regional climatology that students can consult for detailed accounts of regional variations in climate over the world (see for example Kendrew, 1947; Trewartha, 1961; Miller, 1965). In the remainder of this chapter we will consider in broad terms the major distinguishing features of tropical and temperate climates as well as the factors that account for the observed regional variations in the climates of the tropics and the temperate region. These will be illustrated with suitable climatic graphs and diagram from selected representative stations.

On tropical and temperate climates

Tropical climates differ in many respects from temperate climates. Perhaps the most important difference lies in the nature of the energy balance. Insolation receipts in the tropics are much higher than those in the temperature region though the seasonal variations are much higher in the latter region. Similarly, air temperatures in the low latitudes are higher and show fewer seasonal variations than those in the temperature region. In fact, as mentioned in Chapter 4, the diurnal variations in temperature in the low latitudes are more dominant than the seasonal variations. In the temperate region, the reverse is the situation as seasonal variations are more important than diurnal variations.

In terms of precipitation characteristics and distribution, precipitation in the tropics in synonymous with rainfall except for some hailstorms. Snow is absent except on some high mountains. Because precipitation in the tropics is mostly convectional in origin, it tends to be more localized in its spatial distribution in contrast to the dominantly cyclonic precipitation of the temperate region which tends to be widespread in its spatial distribution. Violent storms (e.g. hurricanes) are virtually restricted to the low latitudes. Precipitation is also more seasonal in its incidence in the low latitudes. In the temperate region, precipitation is generally more evenly distributed through the year except in the continental interiors. Weather in the temperate region is, however, more changeable than weather in the low latitudes. This has given rise to the popular notion that the temperate region has weather not climate while the low latitudes have climate not weather. The changeability of weather in the temperate region is due to the fact that the area comes under the influence of various air masses with contrasting thermal and humidity characteristics. This situation encourages the formation of frontal depressions and their accompanying anticyclones which dominate the weather and climate of this region. In contrast, the air masses in the tropics are fewer in number and are rather similar particularly in their thermal properties. Frontal depressions, and the associated sequence of weather types that they bring are therefore absent.

Apart from the above, the climates of the tropics and the temperate region are subject to similar controls. Examples of such common controls include ocean currents, topographic effects, and continentality effects arising from the pattern of distribution of land and water surfaces. In both the tropics and the temperate region, there are differences in the climates of maritime locations and continental locations, between lowlands and highlands and between coastal areas washed by cool currents and those washed by warm currents. Let us now consider the major characteristics, types, and distribution of tropical and temperate climates.

Regional variations in temperate climates

Every climatic classification scheme recongizes more climatic types in the temperate region than in the tropics. Temperate climates generally prevail poleward of latitudes 30° north and South of the equator while tropical climates are found between the two 30° parallels. As mentioned earlier, weather and climate in the temperate region are dominated by pressure systems which control the daily and seasonal variability of the weather. There are, however, marked spatial variations in the climate of the temperate region arising from the effects of the following factors:

1 location relative to the oceans;
2 latitudinal location which determines the radiation regime and the length and severity of the cold season;
3 topography which influence both temperature and precipitation;
4 degree of influence of ocean currents, warm or cold.

Temperate climates vary from the warm type through the cool type to the cold type. Using precipitation amount as a distinguishing characteristic, temperate climates may be moist or dry in varying degrees.

Areas of the world having temperate climates are the following:

1 Europe;
2 most parts of North America;
3 Asia excluding southern and southeastern Asia;
4 New Zealand and the eastern parts of Australia;
5 southern parts of South America;
6 northern and southern fringes of the continent of Africa.

Seasons in the temperate region are determined by temperature rather than precipitation amount as in the tropics. The middle and high latitudes of both hemispheres are influenced by contrasting air masses whose advance and retreat control weather and climate as well as human activities. The major types of temperate climates are briefly described below. Each climatic type is illustrated with the aid of temperature and rainfall graphs for typical stations.

Just outside the tropics on the west coasts of the continents the climate is characterized by a hot dry summer and a mild and rainy winter. This is the *Mediterranean climate* (Fig. 11.3), so called because of its prevalence in the land areas bordering the Mediterranean Ocean. The same type of climate is known as the dry summer subtropical climate. Annual precipitation in this climate ranges from 350 mm on the semi-arid margins to 900 mm towards the marine west coast climate. Mean annual temperatures are lower than those in the tropical climates. The climate is thus subtropical in terms of temperature and latitudinal location. Winter precipitation comes from cyclonic storms while the little summer rainfall consists mainly of scattered showers.

The *humid subtropical climate* (Fig. 11.4) is found in the same latitudes as the Mediterranean climate, but on the eastern sides of the continents. Humid subtropical climate prevails over southeastern USA, eastern coasts of Australia and South Africa, southern and eastern China, northern India and northeastern Argentina. Temperatures are similar to those in the Mediterranean climate but the relative humidities are higher so that summers are more like those in the

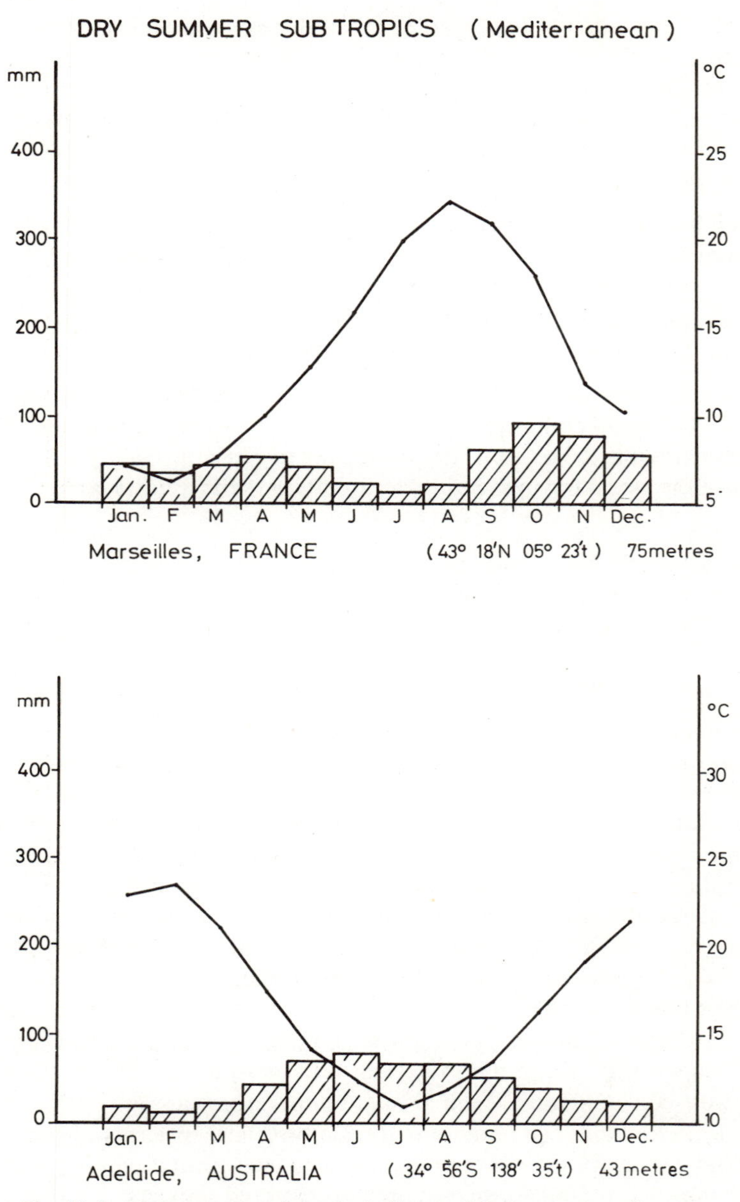

Fig. 11.3. Dry summer subtropical climate (Mediterranean climate)

tropical rainy climates. Annual precipitation totals vary from 750 mm to about 1500 mm. Winters are cool owing to the influence of polar air masses. In most parts of this climatic region temperatures for the winter months vary from 5 to 12 °C.

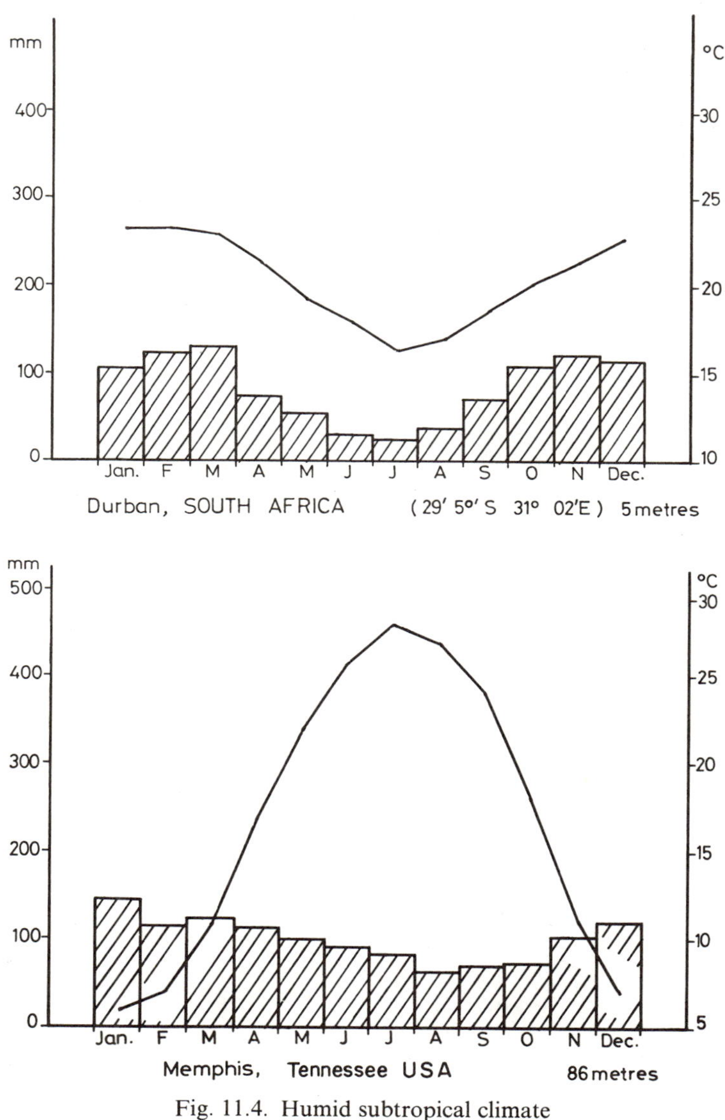

Fig. 11.4. Humid subtropical climate

The *marine west coast climate* (Fig. 11.5) is found on the west coasts of continents poleward from the dry summer subtropics. This type of climate prevails over the British Isles and northwest Europe, the west coast of North America, southern Chile, southeastern Australia, and New Zealand. Mean annual temperatures vary from 7 to 13 °C. The annual precipitation total varies from less than 500 to over 2500 mm with most of it occurring in winter. In summer

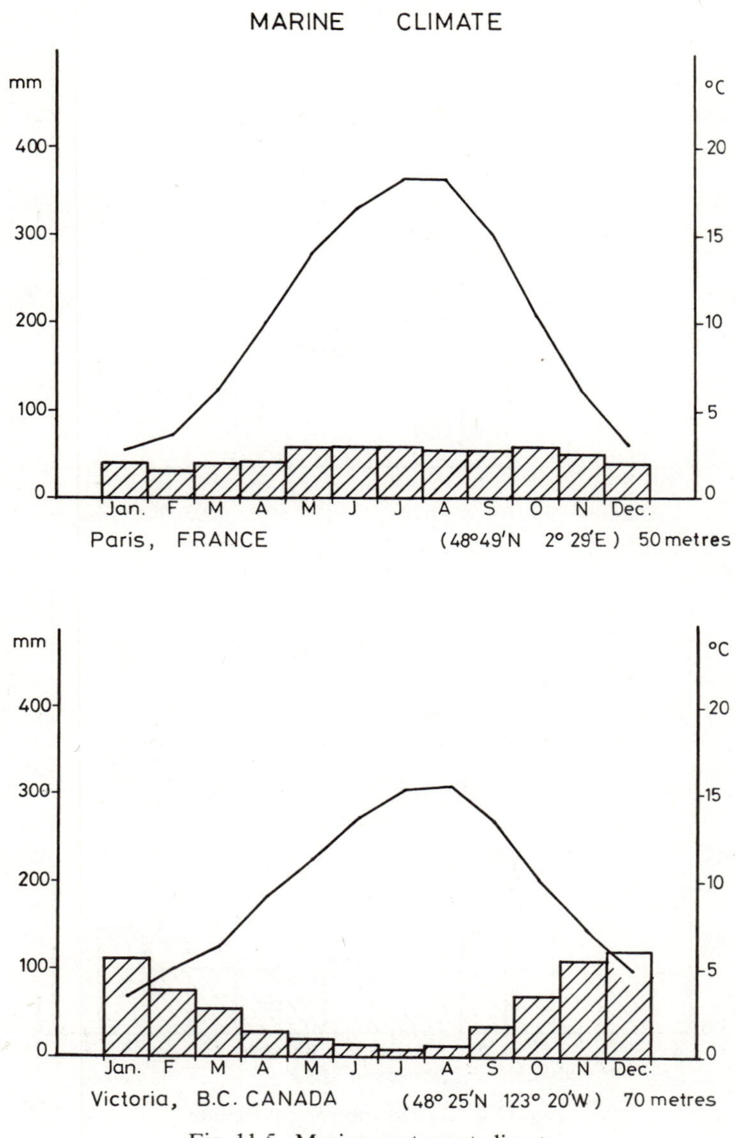

Fig. 11.5. Marine west coast climate

potential evapotranspiration generally exceeds rainfall. Precipitation is mostly cyclonic in origin although thunderstorms are common in summer. Owing to the influence of warm ocean currents, temperatures are mild for the latitude in winter and cool in summer.

Humid continental climates (Fig. 11.6) occur only in the northern hemisphere since the southern hemisphere has mostly water surface at these latitudes. The

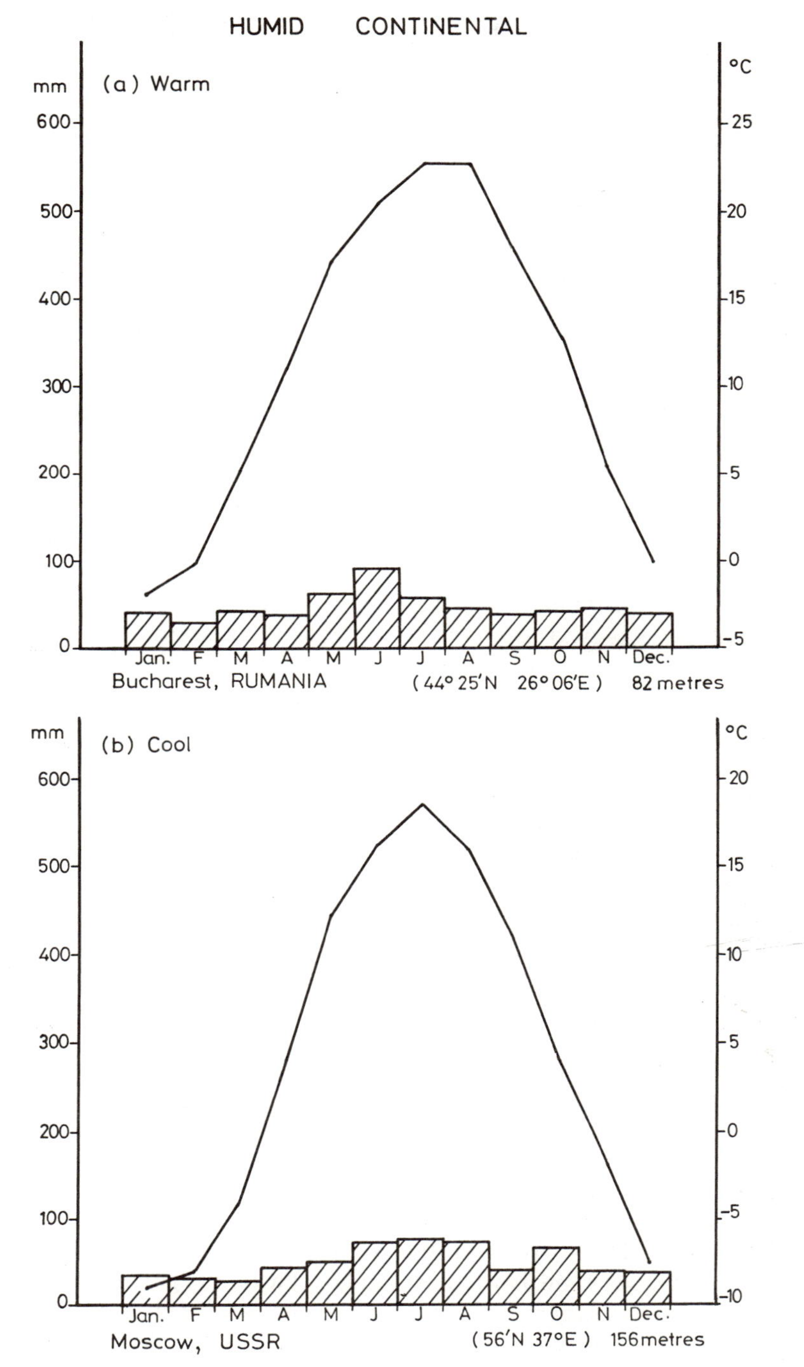

Fig. 11.6. Humid continental climates

climate is characterized by a great annual range of temperature. The winter is cold while the summer is hot (warm summer subtype) or cool (cool summer subtypes). Annual precipitation in the warm summer subtype varies from 500 to 1250 mm, the amount decreasing both poleward and towards continental interiors. Spring or summer is the season of maximum rainfall. Though there are occasional frontal systems, most of the warm season precipitation is due to convective showers. In the cool summer subtype, annual precipitation is less and varies from 350 to 700 mm. Again the maximum precipitation occurs generally in summer or autumn. In winter more snow falls than in the warm summer subtype where winter is less severe.

The *mid-latitude semi-arid and arid climates* (Fig. 11.7) have lower mean temperatures than the tropical arid and semi-arid climates but both are characterized by low precipitation. Unlike in the tropics, subsiding air masses are not the main controls of mid-latitudes arid and semi-arid climates. The major control is their location in continental interiors far-removed from the influence of maritime air masses. The annual range of temperature is large with hot summers and cold winters. Precipitation is low and variable from year to year, about 150–400 mm per annum in the semi-arid climate and much less than that in the arid climate. Desert vegetation is characteristic of the mid-latitude desert while steppe grassland predominates in the semi-arid region. Mid-latitude arid and semi-arid climates prevail over southern USSR, the intermontane basins and Great Plains of the western United States, northern China and western and southern parts of Argentina.

Poleward of about latitude 55–60° north and south of the equator the climate is cold being dominated by polar and arctic-type air masses. The short summers are warm with long days while the winters are long and severe with long nights. Winter is therefore the dominant season. Precipitation which is almost always in the form of snow varies with locations but is generally less than 500 mm. Towards the poles, mean monthly temperatures are all below 0 °C and vegetation is totally absent, the ground being covered with snow and ice. Precipitation is virtually nil hence the name cold desert. Fig. 11.8 shows the characteristic features of the three categories of climates of high latitudes namely *subtractic or Taiga, Tundra, and Polar climates.*

Lastly, there are *highland climates* (Fig. 11.9) which are characterized by low temperatures owing to the normal decrease of temperature with elevation at the rate of about 6 °C per 1000 metres. Highland climates show great diversity from place to place according to latitude and elevation. They also exhibit large differences over short horizontal distances because of variations in local relief. Areas designated as having highland climates are the mountain chains and highlands of middle and low latitudes such as the Rockies and Cascade–Sierra Nevada of North America, the Andes in South America, the Alps in Europe, the Himalayas and Tibet in Asia and the Eastern Highlands of Africa.

Highland climates vary in their temperature and rainfall characteristics since they are located in different latitudinal zones and may be close to or far away from oceans. Highland climates have lower temperatures than the surrounding

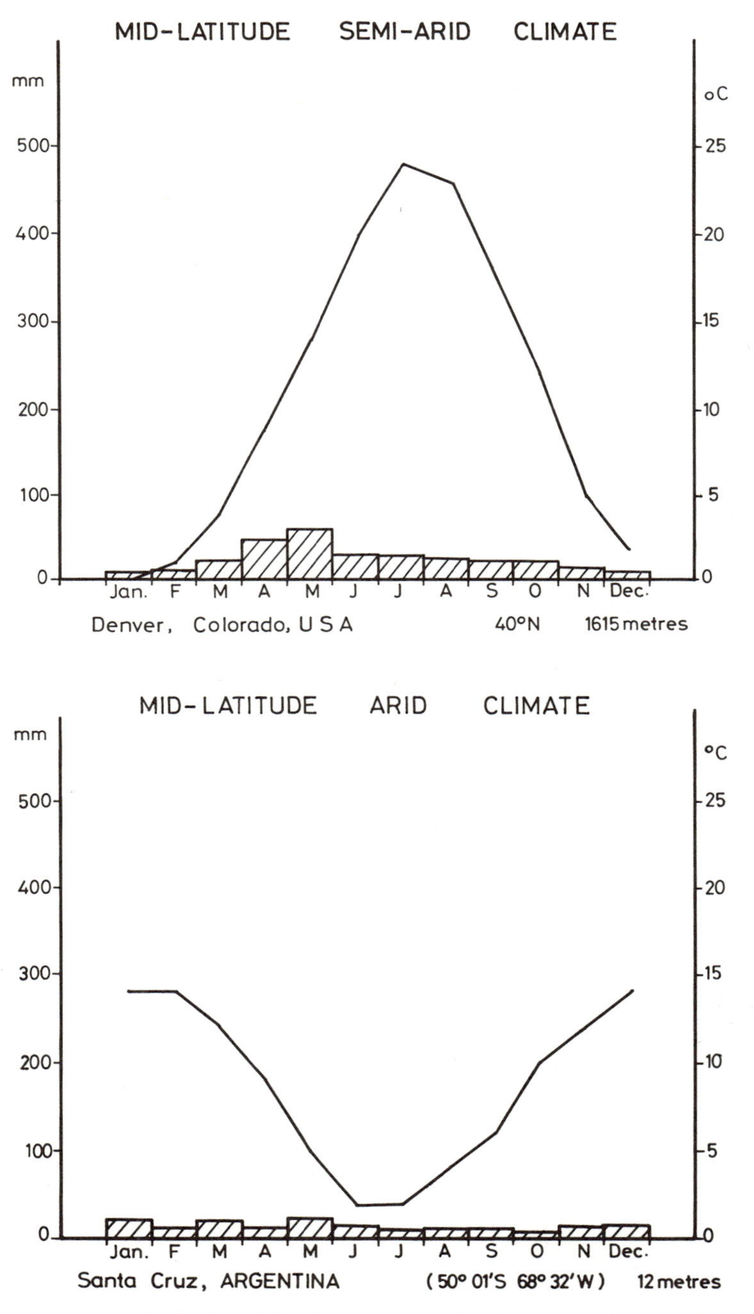

Fig. 11.7. Mid-latitude semi-arid and arid climates

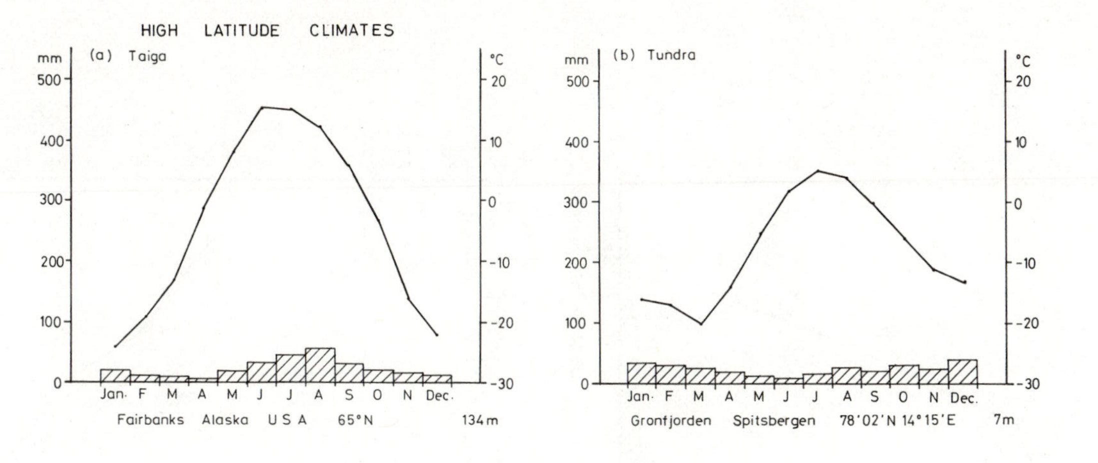

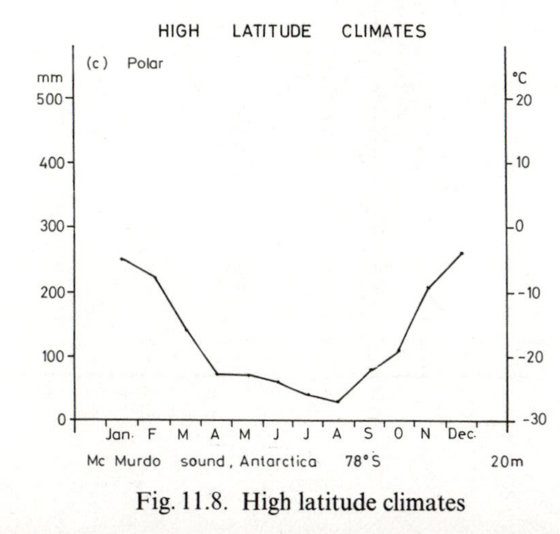

Fig. 11.8. High latitude climates

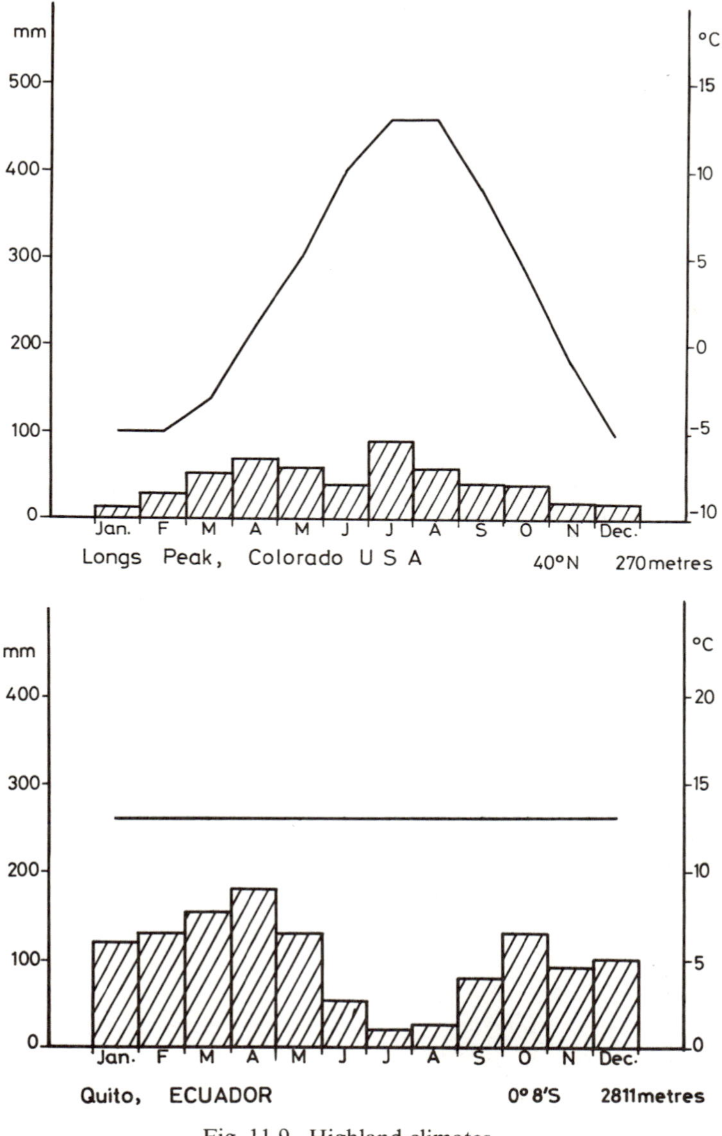

Fig. 11.9. Highland climates

lowland climates. A distinctive feature of highland climates in the tropics is a larger diurnal than annual range of temperature. Also, whereas temperatures in highland climates are lower with increase in elevation, precipitation tends to increase with altitude up to about 3000–5000 metres and then decreases with elevation (Critchfield, 1974). The proportion of precipitation falling as snow also increases with increasing elevation and the snow cover remains for a longer period. Fogginess is generally greater in mountainous areas than on nearby plains or valley.

Regional variations in tropical climates

Tropical climates are characterized by the absence of a cold season and are found in the following major areas:

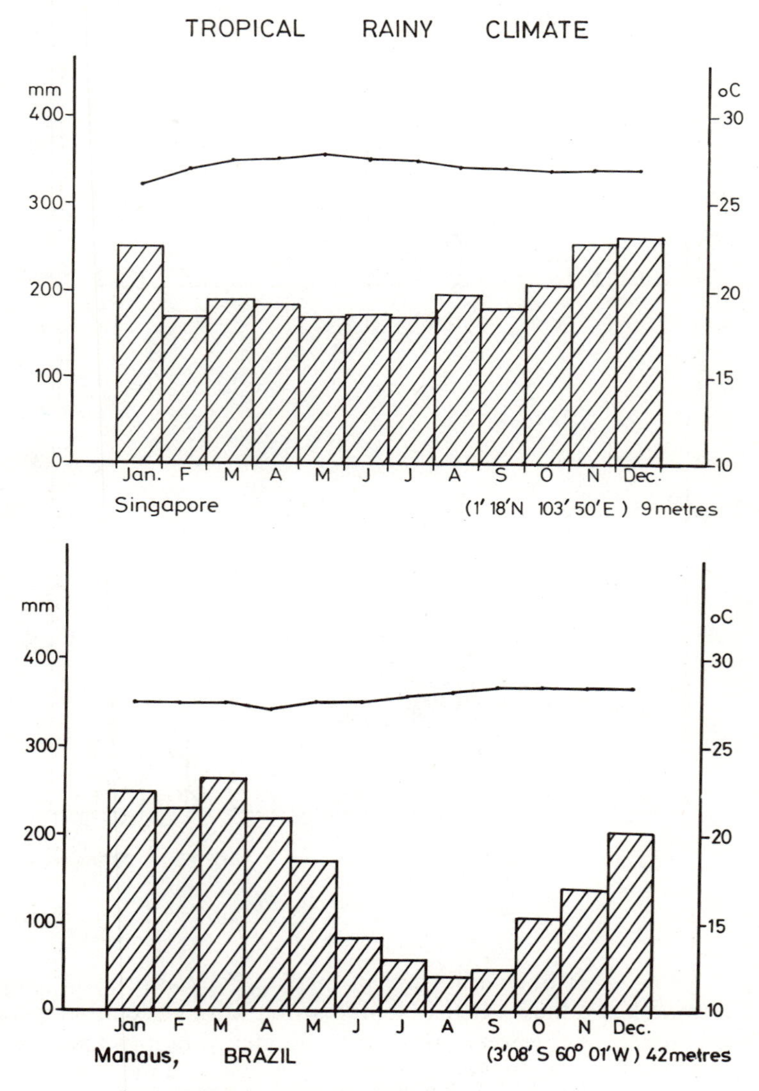

Fig. 11.10. Tropical rainy climate (equatorial climate)

1 the continent of Africa with the exception of the northern and southern fringes;
2 the monsoon Asia covering southern and southeastern Asia and northern Australia;
3 central and the northern parts of South America.

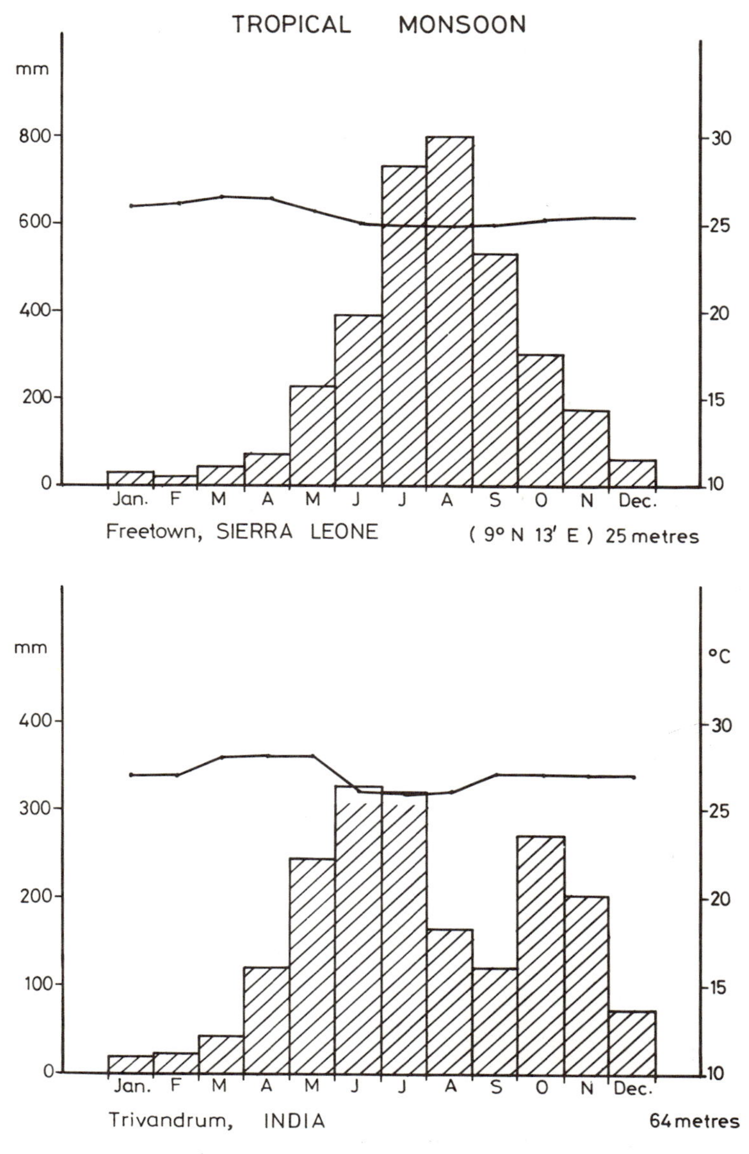

Fig. 11.11. Tropical monsoon climate

Because temperature shows a great deal of uniformity in the tropics, subdivisions of tropical climates are usually based on the amount and distribution of rainfall. Regional variations in tropical climates are therefore due mostly to the two major factors which influence precipitation distribution namely:

1 location relative to the source of moisture supply in the oceans or the intertropical convergence zone; and
2 topography.

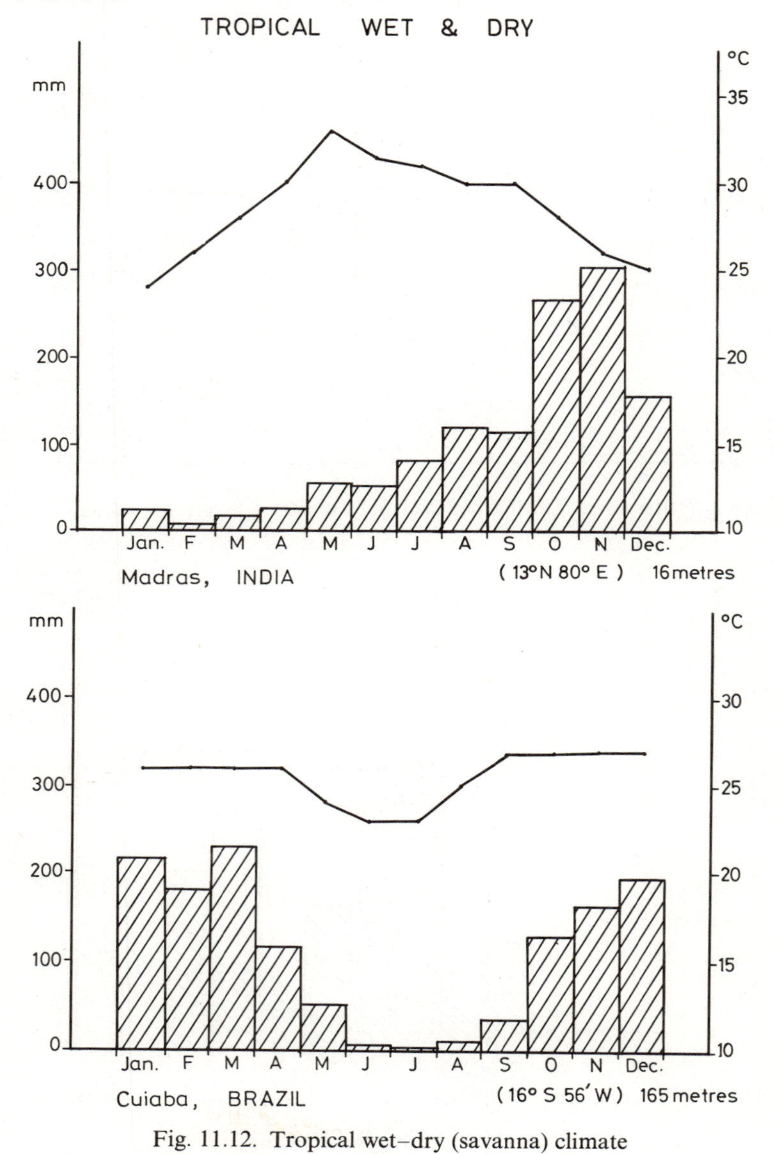

Fig. 11.12. Tropical wet–dry (savanna) climate

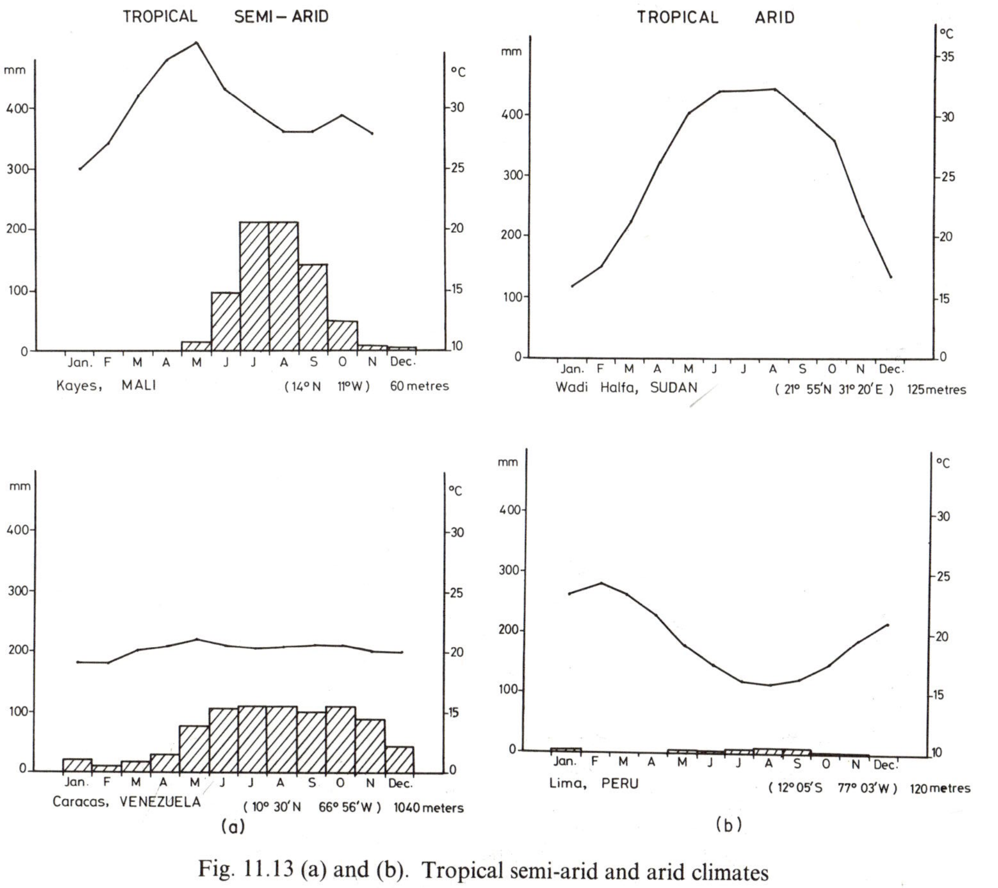

Fig. 11.13 (a) and (b). Tropical semi-arid and arid climates

These factors also partly influence the air temperatures. Marine tropical climates generally have more precipitation but lower temperatures than continental tropical climates. Also marine tropical climates exhibit less diurnal and seasonal variations in temperature than their continental counterparts. Except in a few places mostly around the equatorial zone where rain occurs more or less throughout the year, rainfall over most parts of the tropics is seasonal in occurrence. The rainfall amount generally decreases with increasing continentality while the rainfall seasonality becomes increasingly more pronounced with increasing distance from the source of moisture supply in the ocean. Generally speaking areas of high elevation have better physiological climates than the lowland areas which tend to be hot and humid. The major types of tropical climates are described below with the aid of temperature and rainfall graphs for selected stations.

The *tropical rainy climate* prevails mainly in the lowlands on or near the equator and is therefore also known as the equatorial climate. The climate is characterized by a combination of constantly high temperatures and abundant rainfall well distributed throughout the year (see Fig. 11.10). The climate literally has no seasons with monthly temperatures averaging 25–28 °C. The annual range of temperatures are usually less than 3 °C but the diurnal ranges are much larger and may be 8–10 °C. Rates of evaporation and transpiration are high owing to the constantly high temperatures.

The *monsoon tropical climate* (Fig. 11.11) differs from the tropical rainy climate in that it has a distinct dry season. Annual rainfall totals and temperature conditions are similar to those in the tropical rainy climate but the rainfall regime is similar to that of tropical wet–dry climate. Main areas of monsoon climate are the western coastlands of West Africa, west coasts of India and Burma, and northeastern coast of South America.

The *tropical wet–dry climate* (Fig. 11.12) has alternating wet and dry seasons. Annual rainfall totals are less than in the tropical rainy and monsoon climates. The tropical wet–dry climate is found generally between the average locations of the equatorial low and the subtropical highs in both hemispheres. The dry season here is severe and has a profound effect on vegetation and crops unlike in the tropical monsoon climate. The climate is characterized by savanna vegetation.

Tropical semi-arid and arid climates (Fig. 11.13) are found around latitudes 20–30° north and south of the equator in the zone of the subtropical highs. There is subsidence of air masses resulting in adiabatic heating and low relative humidity. Evaporation rates are very high while rainfall is very low and insufficient to sustain dense vegetative growth. In the tropical semi-arid climate the annual rainfall is at least 250 mm. In the tropical arid climate rainfall amount is not only much lower but is also more erratic. In fact, there is no definite seasonal regime here. Temperatures are rather similar in both arid and semi-arid climates. Maximum temperatures are, however, higher in the arid climate which is reputed to have the highest in the world. Both the annual range of temperatures and the diurnal ranges of temperatures are high, 20–25 °C being common.

References

Ayoade, J. O. (1972). A re-examination of Thornthwaite's moisture index and climatic classification. *Quarterly Meteorological Magazine* (*Lagos*) Vol. 2, No. 4 pp. 190–204.
Ayoade, J. O. (1977). On the use of multivariate techniques in climatic classification and regionalisation. *Arch. Met. Geophys. Biokl. Series B* **24**, 257–267.
Budyko, M. I. (1958). *The Heat Balance of the Earth's Surface.* (translated by N. A. Stepanova) U. S. Department of Commerce, Washington, D. C.
Critchfield, H. J. (1974). *General Climatology* (3rd edn). Prentice-Hall, New York.
Flohn, H. (1950). Neue Auschavvgen über die allgemeina zirulation der Atmosphäre und ihre klimatische Bedeutung. *Erdkunde* **4**, 141–162.
Gabes, E. S. (1972). *Meteorology and Climatology for the Sixth Form and Beyond* (4th edn). Harrap. London.
Johnston, R. J. (1969). Choice in classification: the subjectivity of objective methods. *Annals Assoc. Amer. Geog.* **58**, 575–589.
Kendrew, W. G. (1947). *Climate of the Continents* (3rd edn). Oxford University Press, Oxford.
Miller, A. A. (1965). *Climatology*. Methuen, London.
Steiner, D. (1965). A multivariate statistical approach to climatic regionalisation and classification. *Tijdschr. K. ned aardrijksk. Geroot* **82**, 329–347.
Strahler, A. N. (1969). *Physical Geography* (3rd edn). John Wiley, New York.
Terjung, W. H. and Louie, S. (1972). Energy input output climates of the world: a preliminary attempt. *Arch. Met. Geoph. Bickl. Series B* **20**, 129–166.
Thornthwaite, C. W. (1948). An approach towards a rational classification of climate, *Geog. Rev.* **38**, 55–94.
Thornthwaite, C. W. and Mather, J. R. (1955). *The water balence* Publications in Climatology Vol. 8 No. 1 Laboratory of Climatology, Centerton, New Jersey.
Trewartha, G. T. (1961). *The Earth's Problem Climates.* University of Wisconsin Press, Wisconsin.
Trewartha, G. T. (1968). *Introduction to Climate.* McGraw-Hill, New York.

CHAPTER 12

Climate and Agriculture

Introduction

In spite of recent technological and scientific advances, weather is still the most important variable in agricultural production. The climatic factor affects agriculture and determines the adequacy of food supplies in two major ways. One is through weather hazards to crops and the other is through the control exercised by climate on the type of agriculture feasible or viable in a given area. Climatic parameters have an influence on all stages of the agricultural production chain including land preparation, sowing, crop growth and management, harvesting, storage, transport, and marketing.

Climate–crop relations

Any agricultural system is a man-made ecosystem that depends on climate to function just like the natural ecosystem. The main climatic elements that affect crop production are the same as those influencing natural vegetation. These include solar radiation, temperature, and moisture. These climatic parameters and others dependent on them largely determine the global distribution of crops and livestock as well as crop yield and livestock productivity within a given climatic zone. All crops have their climatic limits for economic production. These limits can be extended to some extent by plant breeding and selection and by cultivation methods in respect of crops and by animal breeding and improved animal husbandry in respect of livestock.

Let us now examine some of the ways in which the major climatic elements influence crop growth and yield. Two important points must be emphasized at the outset. The first point is that climatic variables are closely interrelated in their influence of crops. The effect of a given climatic variable is modified by the others. Also, daily, seasonal, or annual variations in the values of the climatic element are of great importance in determining the efficiency of crop growth. The second point is that in considering the climatic environment in which crops grow, the microclimate immediately around the crop is of vital importance. Climatic conditions within the soil where germination takes place and very close to the ground where the crops grow may be quite different from those prevailing in the free air above the crops.

Solar radiation

Solar radiation is of vital importance in agriculture since it is this energy that powers the agricultural system like any other ecosystem. Solar radiation determines the thermal characteristics of the environment, namely the air and soil temperatures, sunshine and day length or the photoperiod. Photosynthesis, the basic process of food manufacture in nature, and photoperiodism, the flowering response to daylight, are both controlled by solar radiation. The maximum amount of plant tissue which can be photosynthesized within a crop depends on the availability of suitable radiation assuming unlimited carbon dioxide, water, and soil nutrients. The process of photosynthesis which is sometimes referred to as primary biological production can be summarized as follows:

$$[CO_2 + H_2O] \rightarrow \left[\begin{array}{l}\text{Light absorbed by} \\ \text{chlorophyll contained} \\ \text{in cells of green plants}\end{array}\right] = \underset{(CH_2O)}{\text{Carbohydrate}} + O_2$$

In photosynthesis the visible rays are the most effective although ultraviolet rays can influence germination and the energy and quality of seeds. Radiation intensity is also an important factor. The optimum light values for normal crop growth and development are generally around 8–20 kilolux. Light values for optimum flowering and fruiting determined for some crops are as follows:

Pea	850–1100 lux
Corn	1400–1800 lux
Barley and wheat	1800–2000 lux
Bean and cucumber	2400 lux
Tomatoes	4000 lux

If there is insufficient radiation the root system of the crop is underdeveloped, the foliage is yellowish and there is a tendency for the stalk to grow at the expense of the foliage. According to Griffiths (1976) the average plant begins to accumulate organic matter at about 0 °C increasing in amount to 25 °C then decreasing to zero at 40 °C. This indicates that too little radiation or too much radiation is harmful to the process of photosynthesis.

In photoperiodism the important solar rays are those between 0.5 and 0.7 micrometres. Some plants are indifferent to day length while others are very sensitive to variations in day length. Also, some plants are light loving (e.g. pine, birch and larch trees) while others love shade (e.g. beech and spruce). This is not to say that shade-loving plants do not require solar radiation. Crops may be classified as short-day or long-day types, depending on the period when they achieve their optimum growth or crop maturity time. Crops which achieve their optimum growth during the period of short days (about 10 hours sunlight) are said to be short-day crops and they include beans, corn, cotton, millet, cucumber, and tomatoes. Crops which achieve their optimum growth during the period of long days (about 14 hours sunlight) are long-day crops and they include barely, clover, mustard, oats, rye, and wheat. As would be expected tropical plants are generally of short-day type while plants that originate in the middle latitudes are long-day types. Vegetables, like cabbage, lettuce, and spinach which all originate in the middle latitudes will have their flowering retarded if grown nearer the equator as the day length will be shorter than what they are accustomed to in the middle latitudes.

Temperature

The temperature of the air and soil affects all the growth processes of plants. All crops have minimal, optimum, and maximal temperature limits for each of their stages of growth. Tropical crops like cocoa and dates require high temperatures thoughout the year. Many crops such as coffee, bananas, and sugar cane are very sensitive to frosts. On the other hand, winter rye has low temperature demands and can withstand freezing temperatures during the long winter period of dormancy. Temperatures below 6 °C are lethal to most plants. The upper lethal temperatures for most plants range from 50 to 60 °C depending on the species, stage of growth, and length of exposure to the high temperature. Low temperatures kill or damage plants. Prolonged chilling of plants at temperatures above freezing retards plant growth and may kill those plants adapted only to warm conditions. Although chilling may not directly kill plant cells, it reduces the flow of water from the roots and so interferes with plant transpiration and nourishment. When temperatures are below freezing the living matter of cells may freeze and cell dehydration may occur. This will eventually kill off the plants.

Generally high temperatures are not as destructive to plants as low temperatures provided the moisture supply is adequate to prevent wilting and the plant is adapted to the climatic region. Excessive heat can destroy the plant protoplasm. It has a desiccating effect on plants and the rapid rates of transpiration may lead to wilting.

Most crops cannot be grown successfully unless the temperature exceeds some critical threshold values. For instance, coconuts and pineapples thrive only when temperatures are always above 21 °C for at least part of the growing season. Citrus fruit, cotton, sugar cane, and rice will not grow well if temperatures are below 15 °C. Many vegetables require temperatures of at least 8 °C. The critical temperature for wheat is 3 °C (Hobbs, 1980). The growing season is determined in the temperate region by temperature conditions. Unlike in the tropics where the rainfall conditions determine the growing season. In the temperate region, the growing season is usually determined in either of the following two ways. First, the growing season can be defined as the period between the last killing frost of spring and the first killing frost of autumn. This is the definition commonly used in the United States. Second, the growing season can be defined as the period for which daily mean temperatures exceed some critical value usually taken as 6 °C. This is the definition used in Britain. However, there is disagreement over whether daily, weekly, or monthly mean temperatures should be employed. There is also the question of which temperatures to use between air and soil temperatures. For most crops soil temperatures are more important than air temperatures because they directly influence seed germination. For potatoes soil temperatures must be between 8 °C and 28 °C, the optimum soil temperature being 18 °C. Cotton seeds require the soil temperature to be at least 10 °C but below 18 °C for germination.

The intensity of the growing season is represented by the accumulation of temperature units above the growth threshold. The unit used is the degree-day which takes into account the amount by which the daily mean temperature exceeds the stated minimum, usually 6 °C. Thus a day of 16 °C would count as 10 degree-days. The concept of degree-days can be used to estimate the suitability of a crop for a given climatic region assuming water is available. Once the threshold temperature for a crop is known, the accumulation of temperature units above this threshold can be calculated from available temperature data. This can also provide an index for comparing the thermal needs of plants as well as for indicating the suitability of a location for a particular crop. It is pertinent to point out, however, that the concept of degree-days or accumulated temperatures is based on the assumptions that:

1 there is only one significant base temperature operative throughout the life of the plant;

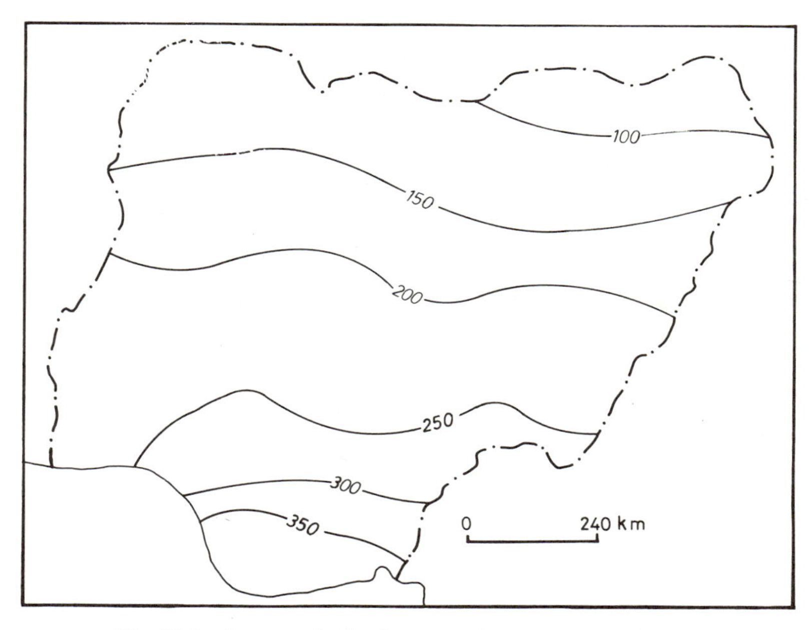

Fig. 12.1. Pattern of effective growth energy over Nigeria

2 day and night temperatures are of equal importance for plant growth;
3 the plant response to temperature is linear over the entire temperature range.

A rather similar index has been computed for Nigeria by Oshodi (1966). The index known as the effective growth energy is based on both temperatures and rainfall. First, a growing month is determined in terms of some critical rainfall amount that must be received. The effective growth energy is then computed as the summation of mean monthly temperatures in excess of 40 °F (4.4 °C) during all the growing months at a given location. The pattern of effective growth energy in °F over Nigeria is shown in Fig. 12.1. The values of effective growth energy decrease from over 350 °F in the southeastern parts of the country to less than 100 °F in the extreme northeastern parts. Oshodi (1966) has also related this energy index to some crops grown in Nigeria as follows:

Crop	Effective growth energy (EGE)
Swamp rice	EGE greater than 350 °F
Oil palm	EGE 250–350 °F
Rubber, cocoa, coconut	EGE 250–300 °F
Rice, maize, beniseed	EGE 200–250 °F
Sorghum	EGE 100–250 °F
Cotton and groundnut	EGE 100–200 °F
Millet	EGE 75–150 °F

Moisture

Water in all its forms plays a vital role in the growth of plants and the production of all crops. It provides the medium by which chemicals and nutrients are carried through the plant. Water is also the main constituent of the physiological plant tissue and a reagent in photosynthesis. Soil moisture is the source of water of importance to crop and the state of soil moisture is controlled by rainfall, the evaporation rate, and soil characteristics. The supply of soil moisture may range from wilting point when no water is available for plant use to field capacity when the soil is fully saturated with moisture but is still well drained. When soil moisture is excessive all the soil pores are completely filled with water and a water logged condition prevails. In such a situation free movement of air within the soil is impeded and compounds toxic to the roots of plants may be formed. At the other extreme is the condition of drought in which the amount of water required for evapotranspiration exceeds the amount available in the soil. Unless this water deficit is made good by rainfall or irrigation the plant will begin to wilt and die. Thus, like extremely low and high temperatures, too much or too little water is not good for agriculture.

The role of moisture in agriculture is even more spectacular in the tropics

Table 12.1 Rainfall requirements of selected food and commercial crops grown in West Africa

Crops	Mean annual rainfall
Yams	At least 1250 mm
Kolanut	At least 1250 mm
Groundnut	500–1000 mm
Beniseed and soya beans	1250–1500 mm
Oil palm	1500–3000 mm
Cocoa	1250–2000 mm
Rubber	2000–2500 mm
Cotton	625–1250 mm

where because of relatively high temperatures throughout the year, the rates of evapotranspiration are constantly high. On the other hand, rainfall is highly seasonal over most parts of the tropics. Because temperatures are high enough throughout the year to ensure the growth of crops over most parts of the tropics with the exception of a few mountainous areas, the growing season, unlike in the temperate region, is determined by the availability of rainfall. Some have used critical threshold values of rainfall to delimit the length of the growing season. For instance, the number of months having at least 100 mm rainfall has been regarded as constituting the growing season. A more useful parameter is the number of months when rainfall is enough to satisfy the evapotranspiration needs of plants. It has been shown in Nigeria by Kassam *et al.* (1976) that crop water requirements are usually fully met when rainfall is only half the value of potential evapotranspiration. The growing season in northern Nigeria has, therefore, been defined by them as the period during which rainfall is at least half the amount of potential evapotranspiration. The calculation of the length of the growing season is a hazardous task because of the variability of rainfall about the average value. Also, the beginning and end of the rains vary sometimes markedly from one year to another. For these reasons, some studies of crop water requirements in the tropics have emphasized the reliability aspect of critical rainfall amounts required for the growth of particular crops (see for example Manning, 1956). The rainfall requirements of some food and commercial crops widely grown in West Africa are shown in Table 12.1. Areas with high probability (say about 80%) of the critical rainfall amounts being received in a given year may be considered for the growth of the relevant crops provided the soils are suitable.

Wind

Wind, air in motion, is another climatic parameter that affects agriculture. On the positive side, wind is an effective agent for the dispersal of plants. The carbon dioxide intake of plants and rates of transpiration tend to increase with increasing wind speed up to a certain level. On the negative side, wind may cause physical damage to crops. By encouraging a high rate of transpiration high winds can also result in plant desiccation. Winds help in the transport of pollen and seeds of undesirable plants such as weeds. Wind erosion can damage good agricultural land by removing the top soil as shown in Oklahoma and Kansas, states of the United States, during the 'dust bowl' period of the drought years of the 1930s. Also high velocity winds in relatively dry areas or during the dry season in subhumid regions can increase the risk of forest fires that can damage farm crops.

It is clear from the above that climatic conditions not only influence the growth and development of plants but also largely determine crop yield. Where optimum climatic conditions prevail and the soil is good, crop yields will be high. At the same time extreme climatic conditions constitute grave hazards to agriculture and have to be controlled. It is to these weather hazards that we will now direct our attention.

Weather hazards in agriculture

The growth of crops is not only dependent on weather conditions but crops are subject to a number of weather hazards until they are harvested. The major weather phenomena that constitute hazards to agriculture are frost, drought, hailstones, and high winds. The nature of these phenomena, the hazards they pose to agriculture, and ways of managing them will now be described.

Frost

Frost is said to occur if the temperature of the air in contact with the ground (ground frost) or at screen level (air frost) is below 0 °C. It is ground frost that is particularly important in agriculture. In weather forecasts, the term 'ground frost' signifies a grass minimum temperature below 0 °C. There are two main genetic types of frost: radiation frost and advection or air mass frost. Radiation frost results from the rapid cooling of the air layer close to the ground following large terrestrial radiation losses on clear, calm nights. Advection or air mass frost occurs when an area is invaded by a cold air mass. Consequently, advection frost affects a large area when it occurs, whereas radiation frost tends to be spotty in occurrence.

Frosts are largely unknown in the tropics except in isolated mountainous areas. Frosts are, on the other hand, common in the temperate region and in the subtropical areas which suffer occasional incursions of cold air masses. Because the temperature of air masses cannot be controlled on a large scale, not much can be done to control or forestall the hazard posed by advection frost. Prevention of crop damage by radiation frost is, however, more feasible. Preventive measures are based upon knowledge of the conditions which favour the occurrence of radiation frost. The conditions are:

1 cool, stable air mass;
2 cloudless sky to allow loss of heat radiation from the earth's surface;

lizers. Weeds must be controlled since they accelerate water loss by transpiration at the expense of crops. In subhumid and semiarid environments, the technique of dry farming is commonly practised. This involves the use of two or three years rainfall to raise one year's crop. During the first year or two, the field is allowed to lie fallow. It is only cultivated to kill weeds and create a soil structure that will retain as much moisture as possible (Critchfield, 1974).

The most effective method of combating drought is through provision of water artificially or by irrigation. Artificial stimulation of rainfall is at present an insignificant method of combating drought. On the other hand, irrigation is a widespread and common method of providing all or part of the water needs of crops. In an arid environment agriculture is possible only with irrigation. In semi-arid and subhumid areas irrigation increases crop yield and the length of the growing season and this makes possible the cultivation of a greater variety of crops. In a humid region, irrigation helps to combat the effect of drought and increase crop yield. But irrigation has its own problems. The major problems of irrigation are:

1 availability of water, surface or underground; and
2 cost of exploitation and getting it to the fields.

There is also a need for judicious application of irrigation water to the crops in the field. The water needs of crops at various stages of their growth must be carefully assessed. While underapplication of water is undesirable, overapplication is also not desirable as it can reduce crop yield and create other problems. Excessive irrigation can:

1 reduce the uptake of nutrients by the plant because of dilution;
2 cause leaching of nutrients out of the reach of crops;
3 oversaturate the soil with moisture so that lack of oxygen becomes a problem.

Above all, irrigation water is often limited and expensive so that overapplication does not make sense economically or ecologically.

Hailstones

Hailstones can physically damage young crops in the field and so constitute a major hazard to agriculture wherever they occur frequently. Hailstones are hard pellets of ice of variable size and shape that fall from cumulonimbus cloud. Hail is, therefore, a solid form of precipitation. Hailstones occur both in the temperate region and in the tropics. Crop losses due to hailstones may be considerable. Hence, various measures are usually taken to minimize or prevent crop damage from hailstones. These have invariably involved the seeding of cloud with silver iodide released from artillery shells or aircraft. The aim of seeding is to create more small ice particles and more but smaller hailstones which are less damaging to crops. Cloud seeding for the purpose of suppressing hail is very common in Russia and has been carried out mostly on an experimental basis in parts of East Africa, notably Kenya.

Wind

Wind transports moisture and heat in the atmosphere and therefore has some effect on crop production. Wind also influences rates of evapotranspiration and directly exerts pressure on crops along its path. Wind may, therefore, constitute a hazard to agriculture in one or more of the following ways.

1 Wind can mechanically damage crops if the speed is too much as the pressure exerted on crops along its path increases with increasing speed.
2 Hot wind will encourage high rates of evapotranspiration and may, therefore, cause desiccation in crops. Besides, the risk of fire is considerably increased.
3 Wind may encourage soil erosion when the speed exceeds some critical threshold value for a given soil environment. Crops may be buried by windblown sand or dust while the stems and leaves of the tall crops suffer abrasion by sand particles.

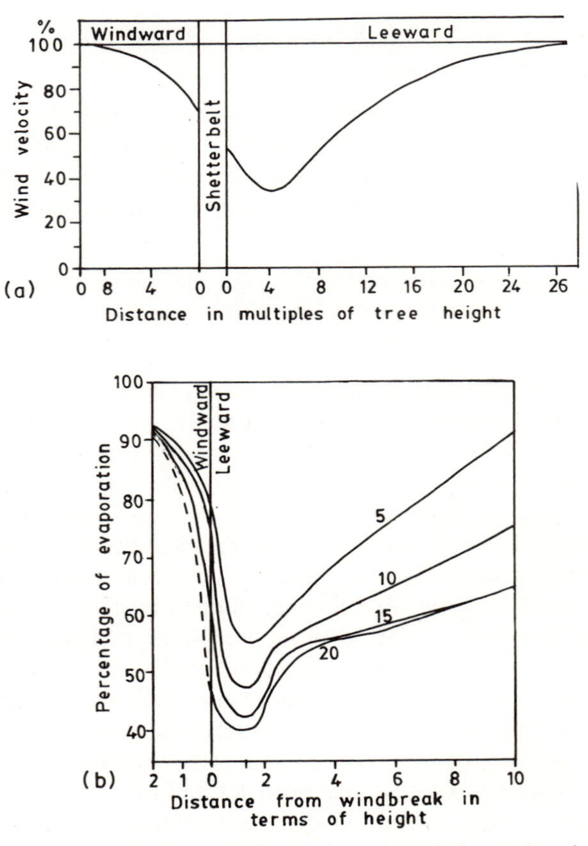

Fig. 12.2. Effect of windbreaks on: (a) wind speed and (b) rate of evapotranspiration

4 Wind may speed the chilling of plants under conditions of low temperature just like it speeds the desiccation of plants under conditions of high temperatures.

Crop damage by winds may be minimized or prevented by the use of windbreaks (shelter belts). These are natural (e.g. trees, shrubs, or hedges) or artificial (e.g. walls, fences) barriers to wind flow to shelter animals or crops. The degree of shelter from wind provided by a windbreak depends on many factors including the following:

1 the height of the barrier;
2 the lateral extent of the barrier;
3 the permeability of the barrier (i.e. the ease with which air can flow through the barrier);
4 the angle of incidence of wind to the barrier.

The effects of windbreaks on wind speed and rates of evaporation are shown in Fig. 12.2. There is a reduction in both wind speed and rates of evaporation before the windbreak is reached. The decrease becomes sharper immediately downwind of the barrier and thereafter becomes less noticeable until wind speed and evaporation rates reach their prebarrier levels. Apart from influencing wind speed and rates of evaporation, air temperature and humidity, soil temperature and moisture are also altered over the area affected by the presence of a windbreak.

Weather aspects of crop pests and diseases

In addition to direct weather hazards discussed earlier, agricultural production also suffers from periodic outbreaks of pests and diseases which are weather dependent. Crop losses caused by pests and diseases may be considerable, particularly in the humid and subhumid tropics. Crop losses due to insect pests in Nigeria have, for instance, been estimated to be of the order of 50–60%. The pests attack the crops in the field and after harvest in barns where they are stored. Some insects also act as vectors of disease-causing germs and should therefore be considered as deadly as the germs they carry.

The periodic or seasonal nature of outbreaks of many crops pests and diseases suggest that weather conditions play an important role. These epidemics are often weather dependent either in terms of local weather conditions being favourable for their growth and development or in terms of the prevailing winds helping to import airborne germs or spores into given areas. Also, some disease-causing viruses are transmitted or spread by insects (e.g. aphids and leaf hoppers) so that the weather conditions suitable for the propagation of these vectors are those which favour the transmission of such diseases.

In the humid tropics and the temperate region, temperature appears to be the critical factor influencing the outbreaks of crop pests and diseases. In arid, semi-arid, and subhumid environments rainfall is the dominant factor. Whatever the environment, however, the crop microclimate is of fundamental importance in the epidemiology of crop diseases. The optimum temperature for the reproduction of aphids is about 26 °C. A delayed spring is favourable to the outbreak of cornseed maggot while a hot dry spell will terminate such an outbreak. Over most parts of the African savanna, grasshoppers and locusts destroy many farmlands every year. The locusts generally originate in the Sahara desert margins where there is enough moisture for breeding and for vegetative growth to feed the larvae. The locusts fly in swarms southwards with the northeasterly winds during the day when temperatures are between 20 and 40 °C. Locusts find it impossible to hold to a course if the wind speed exceeds 16–20 km/hour.

Other crop enemies like mildew, rusts, scabs, and blights reproduce and spread most rapidly when the weather is warm and very humid. In the cocoa growing areas of southwest Nigeria, it has been established that too much rainfall reduces the number of cocoa pods per tree and increases the degree of infection by the black pod disease (Adejuwon, 1962). Also, in Nigeria the incidence of head mould which attacks sorghum in northern Nigeria has been partly attributed to high atmospheric humidity (Kassam *et al.*, 1976). Spores of fungus diseases are spread by wind and this makes their control rather difficult.

Although the control of crop pests and diseases is the work of the plant pathologists, climate is an important factor in such control and must, therefore, be considered by the plant pathologist. Any agricultural practice that provides an unfavourable climatic environment for a crop pest can control such a pest. In this connection, it is important to emphasize the fact that knowledge of crop microclimate or what is often referred to as crop bioclimatology is of fundamental importance in the control of crop diseases and pests. The climatic conditions which favour the propagation of crop pests and crop disease vectors must be carefully studied and identified. Unfavourable microclimatic conditions around crops can be artificially modified and improved. Where this is not possible and insecticides and fungicides have to be used, knowledge of the climatic conditions prevailing is still relevant. The spraying of crops with chemicals is best done under conditions of light winds when the dispersal of such chemicals will be minimized. In the tropics where atmospheric turbulence and instability are commonplace, such calm conditions occur most often in the mornings and evenings.

Climate and animal husbandry

Climatic conditions influence animal husbandry directly or indirectly in three main ways. First, the availability of feed is highly weather dependent and since domestic animals are highly dependent on the availability of feed, climatic factors which influence the growth of pastures or feed crops exert influence, albeit indirectly, on livestock. Climatic conditions determine the type, quantity, and quality of the available feed. The climatic elements which affect animal feed supply are the same as those which influence plant growth or the spread of insects and diseases that attack grain or forage crops. These include rainfall, temperature, and radiation.

Second, climatic conditions have direct influence on domestic animals as they have effects on their normal body functions. For animals to survive in a given climatic zone they must be physiologically adjusted to that climatic environment. All animals and various breeds of particular animals have their optimum climatic requirements to ensure maximum growth and development. When animals are moved to climatic environments they are not used to they generally fall below minimum economic levels of production even though they may not die off like plants do (Critchfield, 1974).

Finally, climate influences livestock production indirected by determining the types of animals and diseases that would be prevalent in a given environment.

The above three ways in which climate influences animal husbandry will now be briefly elaborated upon. The amount and nutritive quality of pastures are impaired by drought conditions. Consequently, under such conditions there is a reduction in milk production and general loss of weight of the animals. There is also a shortage of water for drinking by the animals. This adds to the discomfort of animals, particularly in the warm season and environments. Livestock numbers are drastically reduced by drought as many animals die of hunger and thirst and others are slaughtered for sale by their owners to prevent further deaths and facilitate rationing available feed amongst fewer livestock. Thousands of livestock perished in the West African sahel drought of 1973–4 in such circumstances. Low temperatures in winter also affect livestock feed. Pasture ceases to grow and farmers have to make do with the pasture already conserved for winter use. There are generally no problems encountered in this provided the winter is of expected intensity and duration. If, however, the winter is very severe or unduly prolonged, farmers may find it difficult to feed their livestock from the conserved pasture.

Animals, like human beings, are directly affected by the weather elements. Extremely low or high temperatures interfere with the physiological functions of animals including man. The productive capacities of animals are generally reduced by high temperatures. Dairy cows tend to produce less milk under hot conditions while beef cattle produce less fat and flesh. This is because these animals reduce their intake of feed under hot conditions. For instance, cattle generally prefer grazing in shaded locations than in the sun and they normally rest during the afternoon when temperatures are high and sunlight is strongest. This is not surprising since animals experience discomfort under extreme heat just as human beings do.

The reproductive capacities of animals also tend to decrease under high temperature conditions. Temperate breeds have been known to show a marked decline in fertility under tropical climates (Critchfield, 1974). Until fairly recently, little attention has been given to the question of the adaptation of livestock to hot climates. We now know that livestock in a hot climate are influenced directly by heat radiation and possibly humidity on the animal itself and indirectly by the effect of heat on the animal's environment (Webster and Wilson, 1966). For any livestock to survive and be productive in the tropics, it must be heat tolerant. This means that the animal must have a high efficiency of energy utilization and allow productive processes to continue at a high level without the production of excessive amounts of heat. Another effect of high temperature is to increase the water requirements of livestock. Some of the increased water intake is used to replace water lost from the body of the animal by evaporation. Heat load on animals is only one of the several factors which determine the productivity or even survival of livestock in the tropics. Others include the quantity and quality of feed and the incidence of animal pests and diseases discussed later.

The effect of extreme cold on animals is also to reduce production. This is because much of the body energy would be used to combat cold. Long exposure to cold may cause frostbite or even death. The water requirements of livestock may also not be satisfied under cold weather. Arrangements should, therefore, be made for adequate shelters to keep the animals warm and for water for their drinking needs during cold periods.

Finally, climate indirectly influences animal production through its effects on the incidence and spread of diseases and pests that directly attack the animals or that influence the quantity and quality of forage crops and feed. The primary influence of precipitation on livestock is through its effect on feed. Range animals cannot move about and graze under heavy snow. Many of the restrictions on productivity and regional distribution of animals result from the effects of pests and diseases. In Nigeria, for instance, the more humid southern parts of the country have been shown to be as suitable as the northern parts for tropical cattle and more suitable for imported temperate cattle because the heat load on cattle is less in the south than in the north (Ojo, 1971). Besides, the south has the advantage of better and more reliable water supplies to meet the water requirements of the animals. The observed distribution of livestock in Nigeria is, however, at variance with the above theoretical findings primarily because most of the south is infested with the tse-tse fly while the north is relatively free of this insect which causes high rates of morbidity and mortality amongst cattle.

The tse-tse fly is found over most parts of Africa except the highland areas and the relatively dry areas. The insect transmits and spreads the germs that cause human and cattle trypanosomiasis. The tse-tse fly tends to exist in tree canopies where transpiration and shade maintain a combination of high humidity and moderate temperatures necessary for its growth and production. In the humid and forested areas of Africa only a few dwarf breeds of cattle resistant to trypanosomiasis are kept. Most of Africa's cattle population are to be found in the tse-tse fly free zones of the African savanna.

Climate and agricultural development planning

It is clear from the foregoing that climate directly and indirectly influences agricultural production. Climate must, therefore, be taken into consideration in the planning of agricultural operations and in the planning of agricultural development in general. The relationship between the atmosphere and the soil–plant system is a complex one which can be simplified and generalized in water balance studies (see Thornthwaite and Mather, 1957).

The term 'water balance' refers to a quantitative expression of the hydrological cycle and its various components over a specified area and period of time. The water balance equation is of the general form

$$P - Q - E \pm \Delta S = 0 \qquad (12.1)$$

where P is precipitation, Q is streamflow, E is evapotranspiration and ΔS represents changes in ground water and soil moisture storage. Assumptions and simplifications are adopted in water balance computations. A common assumption is that all precipitation infiltrates and therefore there is no surface runoff until the soil moisture storage capacity is exceeded. The moisture holding capacity of soil is also commonly assumed to be of a particular value usually 250 mm, even though the value varies with soil and plant types.

In water balance computation or measurement it is also commonly assumed that the rates of actual evapotranspiration can be computed as a function of those of potential evapotranspiration if the available soil moisture held in the soil is known. There are, however, disagreements amongst various authorities as to whether rates of evapotranspiration decrease as the soil moisture decreases. There are three broad viewpoints. The first is that evapotranspiration occurs at the potential rate until all the moisture in the soil is evaporated and transpired. The second is that evapotranspiration occurs at the potential rate only when the soil is at field capacity. Once a soil moisture deficit occurs the rates of evaporation decrease in direct proportion to the amount of soil moisture deficit. The third viewpoint is that evapotranspiration occurs at the potential rate until a certain proportion (ranging from 30 to 70%) of the soil moisture has been evaporated or transpired. Thereafter the rates of evapotranspiration decrease with increasing soil moisture deficit. The implications of these and other factors and assumptions on the results of water balance computations have recently been discussed at some length by Ayoade (1976). Here we are more concerned with the applications of water balance studies in agriculture. As enumerated by Jackson (1977), water balance studies may be, and have been, applied in the following ways:

1. to provide a general overview of the water conditions over an area in terms of amounts of precipitation, actual and potential evapotranspiration, soil moisture storage, and change,
2. as part of model to investigate rainfall–runoff relationships with a view to predicting runoff from rainfall data;
3. to assess the suitability of an area for given crops through analysis of the growing season, the water requirements of the crops, and the quantity and frequency of possible irrigation requirements among, other things;
4. to examine the relationship between climate and crop yield;
5. to asses man's impact on the hydrological environment.

All the terms of the water balance equation are of agricultural importance, particularly precipitation and evapotranspiration. These two elements largely determine runoff and the state of soil moisture storage. Precipitation represents the gross water input while potential evapotranspiration gives an indication of the water requirements of crops. The characteristics of rainfall—its amount, intensity, duration, and probability of occurrance—are all of agricultural importance. In the low latitudes, the beginning of the rains must be known and taken into account in planting crops. If crops are planted too early they are destroyed. If they are planted too late they fail to grow at the optimum rate and their growth may even be terminated when the rains stop. For these reasons, the problem of defining and forecasting the beginning and end of the rainy season in the tropics has engaged the attention of many workers (see for example Manning, 1956; Kenworthy and Glover, 1958; Gregory, 1964; Walter, 1967). The reliability of the rains within the rainy season is also of considerable importance in agriculture. The reliability of rainfall amounts critical for the growth of specified crops over a given area must be assessed before such crops are introduced. Otherwise persistent crop failures arising from the unreliability of rainfall will make the cultivation of such crops uneconomic and may lead to famine and misery among peasants. Such studies of rainfall reliability and applications in agriculture have been carried out in East Africa by Manning (1956) and Kenworthy and Glover (1958) among others. Similar studies need to be carried out in other parts of the tropics, particularly monsoon Asia and the West African sahel that recently experienced a disastrous drought. Such studies of crop–climate relations will enable agriculturists to fit crops to climate and avoid the unpleasant consequences of agricultural planning and development not based on a sound knowledge of the climate.

References.

Adejuwon, J. O. (1962). Crop–climate relationship: the example of cocoa in Western Nigeria. *Nigerian Geographical Journal* **5**, No.1, 21–32.

Ayoade, J. O. (1976). On climatic water budgetting procedures. *Nigerian Geographical Journal* **19**, No. 2, 157–177.

Critchfield, H. J. (1974). *General Climatology* (3rd edn). Prentice-Hall, New Jersey.

Gregory, S. (1964). Annual, seasonal and monthly rainfall over Mozambique. In Steel, R. W. and Prothero, R. W. (eds.). *Geographers and the Tropics*. Longmans, London.

Griffiths, J. F. (1976). *Applied Climatology* (2nd edn). Oxford University Press, London.

Hoobs, J. E. (1980). *Applied Climatology*. Dawson, England.

Jackson, I. J. (1977). *Climate, Water and Agriculture in the Tropics*. Longmans, London.

Kassam, A. H. *et al.* (1976). Improving food crop production in the Sudan savanna zone of Northers Nigeria. *Outlook on Agriculture* **8**, No. 6, 341–347.

Kenworthy, J. M. and Glover, J. (1958). The reliability of the main rains in Kenya. *E. Afric. Agric. J* **23**, 267–272.

Manning, H. L. (1956). The statistical assessment of rainfall probability and its application to Uganda agriculture. *Proceedings Royal Society Series B* **144**, 460–480.

Ojo, S. O. (1971). Bovine energy balance climatology and livestock potential in Nigeria. *Agricultural Meteorology* **8**, 353–369.

Oshodi, F. R. (1966). *Bioclimatological Studies of Nigerian Crops*. Nigerian Meteorological Services, Lagos.

Thornthwaite, C. W. and Mather, J. R. (1957). Instructions and tables for computing potential evapotranspiration and water balance. *Publications in Climatology* **X**, No. 3.

Walter, M. W. (1967). The length of the rainy season in Nigeria. *Nigerian Geographical Journal* **10**, No. 2, 123–128.

Webster, C. C. and Wilson, P. N. (1966). *Agriculture in the Tropics*. Longmans, London.

CHAPTER 13

Climate and Man

Introduction

Climate influences man in diverse ways. Man in turn influences climate through his various activities. Until recently emphasis has been on the control exercised by climate on man and his activities. With the increase in the population and technological/scientific capabilities of mankind it has become established that man can influence, and in fact, has been influencing climate though mainly on a local scale. This concluding chapter surveys the nature of the relationship between man and climate and emphasizes the need for man to live in harmony with climate.

On man–climate interactions

Climate is perhaps the most important component of the natural environment. It affects geomorphological processes, soil formation processes, and plant growth and development. Organisms including man are influenced by climate. The major essentials of life for mankind namely air, water, food, and shelter are climate dependent. The air we breathe is obtained from the atmosphere; the water we drink originates from precipitation and the food we eat has its origin in photosynthesis—a process that is made possible by sunshine, carbon dioxide, and moisture all of which are attributes of climate. Man's clothing and shelter are also influenced by climate. In fact, the primary purpose of clothing is to protect man against the weather elements and improve his physiological comfort. Fashion and the need for modesty are of secondary importance. Similarly, shelter is meant to protect man against the weather elements and wild animals and to make him comfortable. Man's various economic activities are influenced by climate in varying degrees. Such activities include agriculture, commerce, and industry as well as transport and communication, to mention but a few. The influence of weather and climate on man and his activities may be benevolent or malevolent. In other words, as far as man is concerned, climate may be a curse or a blessing. Societies have of often viewed climate primarily as a hazard and have tended to neglect it as a resource. This need not be the case. It is now generally agreed that climate can be considered a resource with both benevolent and malevolent aspects. The beneficial effects such as rain, sunshine, cloud, and wind in the proper proportions of time, place, and intensity or amount should be wisely harnessed and utilized rather than be considered as free goods to be misused. The malevolent effects such as floods, droughts, storms, and blizzards should be controlled and managed rather than be regarded as unavoidable costs. The management of climatic resources therefore involves the rational use of the benevolent effects of weather and climate and the prevention, avoidance, or minimization of the malevolent effects.

The diverse ways in which climate affects man and his activities and the impact of man and his activities on climate constitute the subject matter of the growing field of applied climatology (see Smith, 1975; Hobbs, 1980). The role of climate in the physiological comfort of man was briefly discussed in Chapter 4. In Chapter 12, the role of climate in agriculture was discussed at length. In this chapter we will consider aspects of the important role climate can play in the economic development of an area. Climate if allowed to operate without any interference whatsoever can aid or deter economic development. To prevent the latter, man interferes to control or manage climate. His attempts at controlling climate will also therefore be discussed. Finally, man through his various activities inadvertently modifies weather and climate, often without realizing it. Such inadvertent modifications of climate and their consequences are all briefly examined here.

The impact of climate on society

Climate and climatic variations exert a tremendous influence on society. The impact of climate and climatic variation on society may be positive (benevolent or desirable) or negative (malevolent or undesirable). Societies have often viewed climate primarily as a hazard and neglect it is a resource. Yet climate is both a hazard and a resource depending on time, location, and the values and type of climatic parameters involved.

The climate/society interface may be thought of in terms of adjustment, that is, the extent and ways in which society functions in a harmonious relationship with its climate. Man and his activities are vulnerable to climatic variations. At the same time, the activities of man in certain locations and over a period of time may lead to diminishing adjustment or increasing maladjustment of man to his climatic environment.

The extent to which a society is susceptible to damage by climatic causes is termed its vulnerability. On the other hand, the ability of a society to 'bounce back' when adversely affected by climatic impacts is termed its resilience. In general, a society is more vulnerable:

1 the more its economic activity depends upon weathersensitive factors of production;

2 the greater the unreliability and variability of certain key climatic variables like precipitation and temperature;
3 the lower its level of reserves of food and other materials;
4 the less developed the capacity of its transport system to move supplies from areas of surplus to areas of deficit;
5 the less the society has planned and prepared to deal with adverse climatic impacts.

Resilience of society in the face of adverse climatic impacts also depends on a number of similar factors. A society is likely to bounce back more effectively if:

1 it has accumulated stocks or reserves of food and other materials;
2 it has spare capacity built into the design of its infrastructures like power and water supplies;
3 it has command of financial and material resources, technology, and transportation with which to combat the impact. Intangible factors like social cohesion, morale and public confidence and trust in governments and social institutions may also be important. Where these are lacking resilience may be reduced considerably.

Studies so far carried out indicate that the capacity of a society to absorb adverse climatic impacts is not a simple linear function of its wealth or degree of development. It has been hypothesized by Burton *et al.* (1978) that the most vulnearable societies to adverse climatic impacts are neither the poorest and least developed nor the wealthiest and most highly developed, but those societies in the process of rapid transition or modernization where the traditional social mechanisms for absorbing and sharing losses among the community have virtually disappeared but have not yet been replaced by the accumulated wealth and response capacities of modern developed societies. The relationship between the degree of vulnerability of society to adverse climatic impacts and its socio-economic development status is still not well understood and therefore requires further study and research.

We may now examine some of the ways in which climate and climatic variations exercise an influence on man and his activities. The essentials of life for mankind on this planet, namely air, water, food, clothing, and shelter, are all weather dependent or weather related.

Human health, energy, and comfort are affected more by climate than by any other element of the physical environment (Critchfield, 1974). The physiological functions of the human body respond to changes in the weather. Certain illnesses are climate induced while several diseases that afflict man show a close correlation with climatic conditions and season in their incidence. The climatic elements which directly affect the physiological functions of the human body include radiation (sunshine), temperature, humidity, wind, and atmospheric pressure. Human physiological comfort is determined mainly by temperature, wind, and humidity (see Chapter 4). Differences in a feeling of comfort amongst individuals exposed to same or similar climatic conditions arise from variations in age, state of health, physical activity, type and amount of clothing, past climatic experience or degree of acclimatization.

Human physical vigour is influenced by temperature, humidity, and wind. Generally, high temperature and humidity tend to decrease physical vigour as well as mental vigour. Very dry air or extremely low temperatures may also impair physical vigour and adversely affect attitude to mental work. Extremely low pressure will result in a decreased supply of oxygen to the brain with consequent decrease in mental vigour or alertness. Moderate fluctuations in weather are generally regarded as stimulating to physical and mental vigour. Weather also appears to influence human emotions and behaviour. For instance, crimes, riots, insanity, and other individual and group emotional outbursts seem to reach their peak at the onset of or during hot unpleasant weather (Critchfield, 1974).

Some extremes of weather directly affect human health. Exteremely high temperatures give rise to the incidence of heat stroke, heat exhaustion, and heat cramps. Extremely low temperatures on the other hand may cause frostbite and aggravate ailments like arthritis, swollen sinuses, and stiff joints. Heat-related ailments like heat exhaustion, heat stroke, and heat cramps occur in the low latitudes particularly during the dry season and frequently occur in the temperate region during the hot summer months. Frostbite is unknown in the low latitudes but commonly occurs in the temperate region during severe winter months. Respiratory diseases are aided by dry, dust-laden air which irritates the respiratory passages when breathed in. In West Africa, the dry dusty harmattan wind from the Sahara is frequently associated with high incidence of respiratory diseases. Intense sunlight as occurs in arid and semi-arid tropics or the sunlight reflected from snow-covered ground in the high latitudes may cause some forms of blindness, headaches, and other discomforts. Although ultraviolet rays help to form vitamin D in the skin and devitalize bacteria and germs, they can also cause sunburn and inflammation of the skin. In fact, ultraviolet rays coupled with intense heat help in causing cataract of the eye (Critchfield, 1974).

Climate also plays some role in the incidence of certain diseases that man suffers from. First, climate affects the resistance of the human body to some diseases. Second, climate influences the growth, propagation, and spread of some disease organisms or their carriers (vectors). Extremely low temperatures generally lower the resistance of the human body to infection. Fog mixed with smoke pollutants (smog) is frequently associated with an increase in respiratory ailments. Similarly, dry dusty air tends to make the respiratory tracts more susceptible to infection. Wind is an important factor in the incidence and spread of hay fever since it controls the transport of the allergens that cause the hay fever.

Some diseases tend to be prevalent in certain climatic zones while many diseases, particularly the communicable ones, tend to follow a seasonal pattern in their incidence. For instance, malaria and yellow fever are tropical diseases because the germs causing the diseases are transmitted by species of mosquitoes that only thrive in tropical climates. In the temperate region, pneumonia and influenza are more common in winter than in summer partly because the respiratory tract is more susceptible to infection in winter. Meningitis is more widespread during the cool harmattan season in northern Nigeria because the transmission of the disease is facilitated by people huddling at night into overcrowded conditions for warmth. In general, infectious diseases spread more rapidly during the cool season among the population compared to the warm season when people engage more in outdoor activities.

The effects of climate on human health are, however, not all negative. Favourable climatic conditions can help the human body in warding off disease and in promoting recovery from illness. Fresh air, mild temperature, moderate relative humidity and sunshine all have therapeutic values (Critchfield, 1974). For instance, fresh air and sunlight will help recovery from tuberculosis. Rickets and some skin diseases do respond to sunshine. Favourable climatic conditions only aid recovery; they are no substitute for proper medical attention, good nutrition, and cleanliness.

Climate influences in varying degree the various economic activities that man engages in. These include agriculture and animal husbandry (see Chapter

Table 13.1 Optimal indoor operating climatic range for selected industries (after Landsberg, 1966)

Industry	Temperature (°C)	Relative humidity (%)
Textile	20–25	60
Wool	20–25	70
Silk	22–25	75
Nylon	29	60
Food		
Milling	18–20	60–80
Bakery	25–27	60–75
Candy	18–20	40–50
Tobacco		
Cigarette production	21–22	60–65
Tobacco curing	13–28	65–75
Tobacco storage	18–20	60–65
Miscellaneous		
Paper manufacturing	20–24	65
Paper storage	16–21	40–50
Printing	20	50
Drug manufacturing	20–24	60–70
Rubber production	22–24	50–70
Cosmetics manufacturing	20	55–60
Cosmetics storage	10–16	50
Photographic film manufacturing	20	60–65

Source: Landsberg, H. (1967) *Physical Climatology*, p. 392.

12), manufacturing industry, commerce, and tourism. Climate also influences utilities, transport, and communication systems, often impairing their efficiency and usability. Activities which are carried out outdoors are extremely weather vulnerable. Such activities include mining, construction, tourism, and various leisure activities such as games and sports. These activities can only be successfully pursued if suitable climatic conditions prevail (see Mather, 1974; Smith, 1975; Hobbs, 1980).

Activities carried out indoors are also vulnerable to climatic variations. Weather conditions influence the performance and durability of machinery. The tropical climatic conditions, for instance, tend to encourage rusting of machinery. Besides, certain industrial processes can best take place under certain temperature and humidity conditions (see Table 13.1). If such conditions do not naturally exist they have to be artificially created at additional cost through air conditioning. The supply of materials used in the manufacturing process may also be weather dependent. For instance, agro-allied industries (mainly food processing or fruit canning industries) depend on good harvest and adequate and efficient transportation to bring in supplies. Some industrial goods are also seasonally required by the people. The output of such goods will therefore be determined by prevailing climatic conditions. Examples of such commodities include deodorant, suntan lotion, cold drinks, and ice creams in summer and raincoats, boots, and fur coats in winter.

Climatic costs in industrial production and management can be estimated by considering the effect of climate on the following, among others (see Mather, 1974):

1 space heating/cooling requirements;
2 water supply provision;
3 air and water pollution abatement efforts;
4 warehousing storage, and transportation of raw or finished products;
5 weathering or deterioration of machinery and stockpiled items like fuels;
6 Health, efficiency and morale of the workers.

Retailing and commerce are favoured by climatic conditions that encourage people to go out for shopping. Precipitation and/or extreme cold tend to reduce sales as many would-be shoppers stay at home. Poor visibility and heavy snowfall will also adversely affect sales as well as supply of goods. This is because transporatation systems will be disrupted under such weather conditions. All forms of transportation—road, rail, air, and water—are vulnerable to weather conditions although in varying degrees. The most vulnerable is perhaps air transport. Adverse weather conditions may cause accidents, delays, diversions, and even outright cancellations of journeys with concomitant inconvenience, loss of revenue, time or even lives in some cases. Poor visibility is the single most important weather hazard to all forms of transportation. Others include heavy precipitation (whether rain or snow), high winds or blizzard, violent storms, and ground frost (for road and rail transport only).

Climate also partly determines the way we build our houses and the way we dress. Type of clothing varies from one culture to another and also from one climatic zone to another. The world has been divided into seven zones according to the clothing requirements that would maintain a comfortable heat balance in a normal human body (see Griffiths, 1976). The zones are as follows:

1 the minimal clothing zone—this covers the humid tropics where mean monthly temperatures are between 20 °C and 30 °C. Light material like light cotton is ideal.
2 long flowing robe clothing zone—this covers the hot deserts with high temperatures and radiation. Clothing must protect the wearer from solar rays, allow evaporative cooling during the day and insulation against night time cooling.
3 the one-layer clothing zone—this covers the subtropical belt with mean monthly temperatures between 10 °C and 20 °C. Wool is ideal with light cotton undergarments.
4 the two-layer clothing zone—this covers the temperate region with cool winter where mean monthly temperatures are between 0 °C and 10 °C. Conditions are often humid but radiation is not excessive. The ideal clothing which is wool with cotton undergarment should allow about 6 mm of air to be trapped between the two layers.
5 the three-layer clothing zone—this covers the cold winter region of the temperate belt where monthly temperatures are between − 10 °C and 0 °C.
6 the four-layer clothing zone—this covers the temperate region with severe cold winters where mean monthly temperatures are between − 20 °C and − 10 °C. In this zone the Eskimo fur garments are thought to be the most efficient in maintaining body comfort.
7 the Arctic zone—this is a region in which a comfortable body heat balance cannot be maintained by clothing alone.

Weather and climatic conditions are relevant factors in the efficient siting of buildings, choice of materials, design, and air conditioning of the structure (Critchfield, 1974). Buildings are generally required for two things, namely protection against weather elements and wild animals and creation of an artificial climate suitable for living accommodation, storage, working in or some other specified purpose. In addition, however, the building must be structurally safe and able to withstand the stresses of the prevailing climate during its anticipated lifetime (Smith, 1975). All the three stages involved in the building industry namely design, construction, and air conditioning (the maintenance of satisfactory indoor climate through thermal and ventilation control) are all weather and climate dependent. Construction is an outdoor activity and is consequently directly affected by prevailing weather conditions. Rain, snow, high winds, temperature extremes all have adverse effects on construction activities. Estimates of the number of workable days for construction purposes are made using information on the above weather variables (see Smith, 1975).

The traditional architectural design of buildings often represents some measure of climatic response. Similarly the choice of building materials for roofs, walls, windows, etc., reflects the need to maintain suitable indoor climate for a given climatic zone. Unfortunately, cross-cultural contacts have led in recent times to the adoption of housing styles and choice of building materials which are not in harmony with the prevailing climatic conditions. This is particularly the case in the low latitudes where temperate region types of buildings are being erected in the name of modernization. Needless to say, such structures are naturally not as comfortable as the traditional ones though they are more elegant and aesthetically pleasing. These structures are only habitable because of elaborate artificial modification of indoor climate through the use of electric fans and air conditioners which are not only costly to purchase but also to maintain and use because they consume high amounts of electrical energy. Because of the unreliability of the power supply in many countries in the tropics, these structures become very uncomfortable when there are power cuts which unfortunately occur rather frequently.

The following are the general objectives that housing layout and design should aim at achieving in various climatic zones to ensure the maintenance of a suitable indoor microclimate at minimum costs (see Olgyay, 1963):

1. *Cool region*: increase heat production; increase radiation absorption; decrease radiation loss; reduce conduction and evaporation loss.
2. *temperate region*: because this region is characterized by overheated (hot) and underheated (cold) periods, there is a need to establish some balance by reducing or promoting on a seasonal basis the heat production, radiation, and convection effects.
3. *hot, arid region*: reduce heat production; promote loss of radiation; reduce convection gain, promote evaporation.
4. *hot, humid region*: reduce heat production; reduce radiation gain, promote evaporation loss.

In various climatic zones around the world, local inhabitants have evolved house types well adapted to the environment. Modern architects and builders can learn a lot from native housing design in various climatic zones. Characteristics of native housing can be improved upon through the use of better technology and/or materials. There are seven regions with characteristic native housing types that have incorporated the fundamental features of living with the climate (see Fig. 13.1). The regions are as follows (see Griffiths, 1976):

1. *hot, humid zone*: houses are raised and are usually of one room depth. The raised house allows for some protection against insects and animals and gives the living area a greater air flow and ventilation. Shutters are used for protection against intense rainfall.
2. *hot dry zone*: protection from intense radiation and blowing sand is of

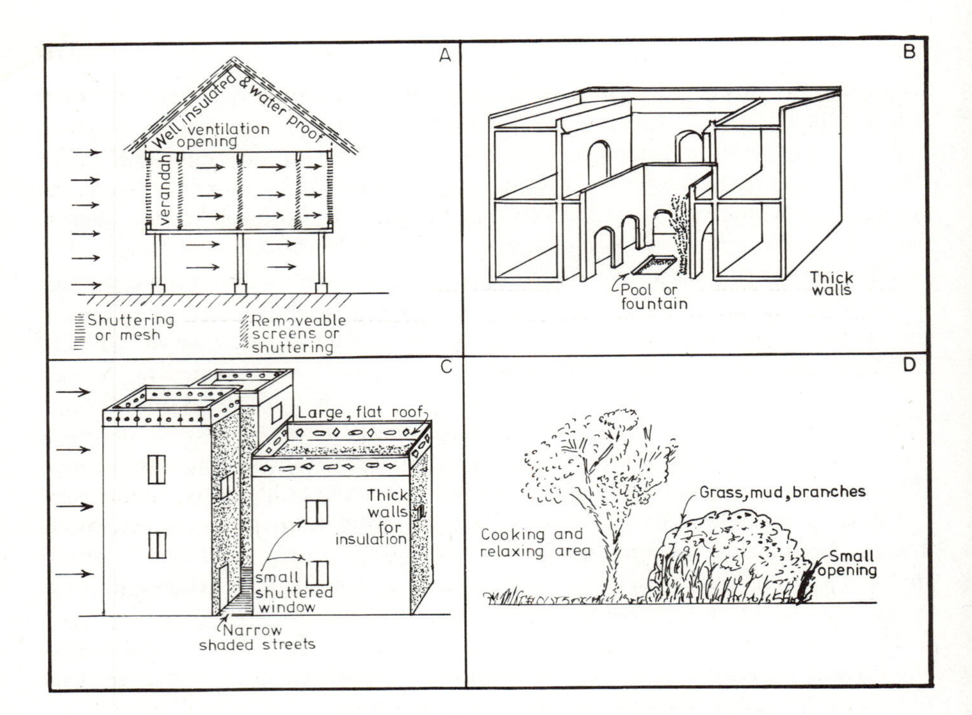

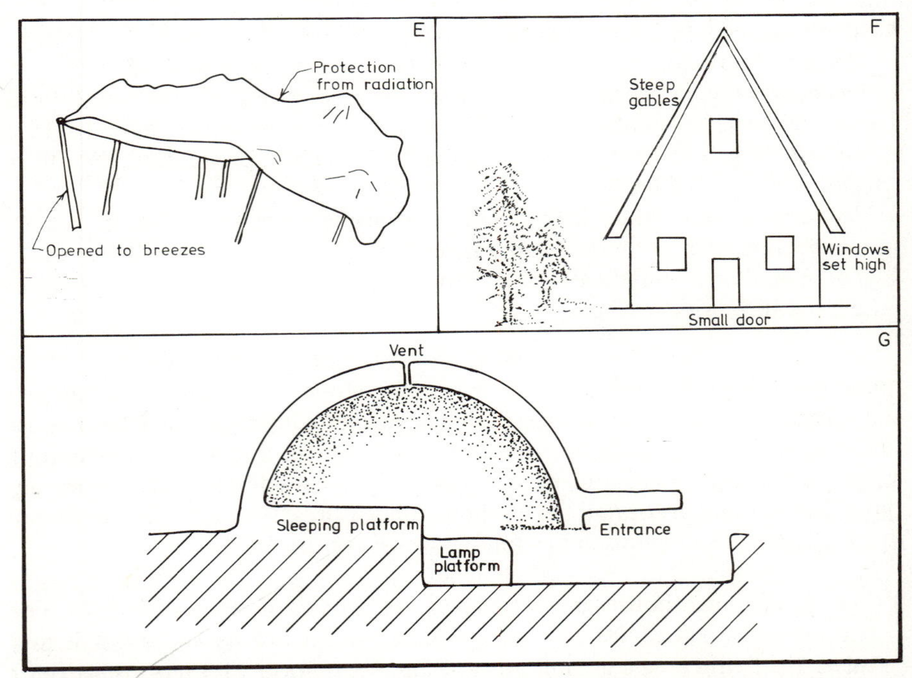

Fig. 13.1. Native house types in various climatic zones A, hot humid zone house; B, Mediterranean zone courtyard house; C, hot dry zone house; D, hot savanna zone house; E, hot desert zone tent; F, cold zone house; G, eskimo igloo (reproduced with permission from Griffiths, 1976)

paramount importance. Dried mud bricks are used for building because they are good insulators. Windows are small and the roofs are flat. The flat roofs are used for relaxation and dinner in the evening. The roofs and walls are painted white to reflect as much radiation as possible. There is also a tendency for the houses to be built close together to shield out as much sunshine as possible.

3 *hot savanna zone*: this zone combines the characteristics of the hot humid zone and the hot dry zone. Dwellings are generally constructed of mud and grass. They are usually erected under the shade of a spreading acacia tree and often surrounded by a thorn fence for security and protection from wild animals.

4 *Mediterranean zone*: this zone has hot dry summers and warm wet winters. Stones and wood are widely used in house construction, the stonework serving as insulating material from the hot rays of the sun particularly in summer. Houses are built with open courtyards with fountains or pools to modify the heat and low humidity of summer.

5 *subtropical desert zone*: temperatures here are very high and vegetation is very sparse. The nomads who inhabit this zone live in tents made of animal skins. The tents can be lifted at the side to take advantage of any breezes.

6 *cool zone*: the problem here is to conserve heat. Houses are usually small to conserve heat. Good insulation is very essential. In the past this was provided by the use of thatch for roof. Double-glazed windows and double doors now help to reduce conductive and draught loss. Since excessive radiation is not a problem, houses are generally designed to take advantage of available sunshine. Unwanted bright sunlight can be controlled by the use of curtains.

7 *cold zone*: in the forested zone wooden huts are common. The roofs are steep to prevent excessive accumulation of snow on them. In the northern parts of this zone temperatures are very low. The Eskimo-developed *igloo* constructed of snow blocks with or without a framework is the ideal type of dwelling. The temperatures within the igloo are artificially increased by the use of oil lamps.

The impact of man on climate

As mentioned earlier, man and climate mutually affect each other. Some of the ways in which weather and climate influence man, his way of life, and economic activities were reviewed in the previous section. In this section, we will consider some of the ways in which man through his various activities exert some influence on local, regional, or even global climate.

Man may influence climate deliberately or inadvertently. Purposeful modification of weather by man is known as *weather control*. Man deliberately attempts to modify weather in order to achieve some objectives. There are several motivations for weather control by man. Man attempts to control weather to achieve one or more of the following:

Table 13.2 Annual average estimated losses from weather hazards in the United States

Weather hazard	Loss
Hurricanes	over $250 million
Lightning-caused fires	$200 million
Hailstones	$300 million
Fog and snow	$66 million

1 to reduce the economic and social losses that result from severe weather events like hurricanes, hailstones, frost, drought, fog, and lightning-caused fires. Table 13.2 shows the annual estimated losses from the effect of adverse weather conditions in the United States of America for which some data are available.
2 to augment the supply of those weather elements that serve as input in the production of particular commodities or services. For instance, the supply of water may be augmented by cloud seeding or by suppressing evaporation from reservoirs by the use of acetyl alcohol, microscopic beads, or shelter belts.
3 to improve health as well as mental and physical efficiency. This involves the creation of an artificial microclimate conducive to the maintenance of physiological comfort in human beings.

From the above, it will be apparent that man's ability to control weather is still very much limited and confined to the local climate. Examples of attempts to control weather by man include cloud seeding to augment precipitation or suppress hail or lightning or to clear fog or modify the structure and movement of hurricanes. The improvement of nighttime microclimates of plants to prevent ground frost (see Chapter 12) and the reduction of evaporation off reservoirs are also examples of purposeful weather modification by man. Everyday man tries to control weather within his home or office by the use of artificial air conditioning systems. These allow him to maintain desirable levels of physiological comfort and to enhance his physical and mental vigour as well as his productivity.

Man may also influence climate inadvertently through his various actions and activities such as urbanization and industrialization, felling of trees (deforestation), farming activities, drainage of marshes or creation of artificial lakes when rivers are dammed to provide water for various uses or for generation of hydroelectric power. Man's greatest impact on climate is evident in urban areas. Here, the actions of man have had such a tremendous impact on climate that the climate prevailing in urban areas is quite distinct in character from that in the surrounding rural areas. In urban areas, the chemical composition of the atmosphere is altered. The thermal and the hydrological properties of the earth's surface as well as its aerodynamic roughness parameters are altered by the processes of urbanization and industrialization. Marshes are drained and natural surfaces are replaced by more impervious surfaces of pavements, tarred roads, and roofs of buildings. As a result of these, both long-wave and short-wave radiation are reduced over urban areas. Temperatures are raised even though the duration of sunshine is reduced. Humidities are reduced but there is some increase in precipitation. Cloud amount in increased. Fogs are thicker, occur more frequently and are more persistent, thus impairing visibility. Turbulence is increased. Strong winds are decelerated and light winds are accelerated as they move into urban areas (see Table 13.3).

The observed climatic changes produced by cities can be explained by reference to the following factors:

1 artificial production of heat from combustion processes, space heating and metabolism.

Table 13.3 Climatic changes produced by cities (after Landsberg, 1970)

Elements	Comparison with rural environs
Pollutants	
Dust particles	10 times more
Sulphur dioxide	5 times more
Carbon dioxide	10 times more
Carbon monoxide	25 times more
Radiation	
Total on horizontal surface	15 to 20% less
Ultraviolet, winter	30% less
Ultraviolet, summer	5% less
Sunshine duration	5–15% less
Cloudiness	
Cloud cover	5–10% more
Fog, winter	100% more
Fog, summer	30% more
Precipitation	
Total amount	5–10% more
Raindays with 5 mm	10% more
Snowfall	5% less
Snow days	14% less
Temperature	
Annual mean	0.5–1.0 °C more
Winter minima	1–2 °C more
Heating degree-days	10% fewer
Relative humidity	
Annual mean	6% less
Winter	2% less
Summer	8% less
Wind speed	
Annual mean	20–30% less
Extreme gusts	10–20% less
Calms	5–20% more

2 heat production as a result of the thermal properties of the city fabric. Buildings, pavements, and roads in cities absorb and store radiation during the day and gradually release this into the atmosphere at night.
3 modification of the chemical composition of the atmosphere as a result of pollutants emitted into the atmosphere from household and factory chimneys and the exhaust pipes of thousands of automobiles plying the city highways. Such man-made pollutants include smoke particles, SO_2, carbon dioxide, carbon monoxide, oxides of nitrogen and chlorine, hydrogen sulphide and ozone. Smoke particles, ozone, and carbon dioxide in particular influence the energy balance of the urban surfaces since these elements can reflect and scatter radiation (smoke particles) or absorb radiation (ozone and carbon dioxide).
4 alteration of the natural vegetated surfaces with artificial surfaces that have different albedo, roughness parameter, and different thermal and hydrological properties.

Two aspects of urban climate are particularly noteworthy because of their wider implications. These are:

1 the increase in temperature over urban areas (the so-called heat island phenomenon), and
2 the pollution of the city air. Urban areas are much warmer than their rural environs particularly at nights.

The urban heat island phenomenon is caused by the following factors:

1 the thermal heat capacity and conductivity of urban surfaces which encourage absorbtion of radiation during the day and its release into the atmosphere at night;
2 the addition of heat from combustion, space heating, and metabolism in the human body;
3 the dryness of city surfaces implies that not much energy will be used for evaporation. Most of the energy will be used for warming the air. The dryness of the city surfaces is due to the removal of surface runoff by urban sewage systems, lack of extensive vegetation cover, and absence of swamps or water pools from which evaporation/transpiration can take place;
4 the decrease in wind flow because of the frictional effect of city structures reduces the exchange of warm city air with that of the cooler rural environ and affects evaporative processes which can produce cooling;
5 the greenhouse effect of the pollution layer over cities also aids the development of the urban heat island phenomenon. There is a reduction in terrestrial infrared radiation to space at night so that energy is conserved within the city atmosphere below the pollution layer.

Studies of the urban heat island phenomenon have been carried out mainly in cities in the temperate region. These studies indicate that the heat island effect is greatest during summer and early autumn on clear calm nights when

Table 13.4 Critical wind speeds for the elimination of the heat island effect in some cities

City	Population	Period of survey	Critical wind speed (m/s)
London, England	8,500,000	1959–61	12
Montreal, Canada	2,000,000	1967–8	11
Bremen, Germany	400,000	1933	8
Hamilton, Canada	300,000	1965–6	6–8
Reading, England	120,000	1951–2	4–7
Kumagaya, Japan	50,000	1956–7	5
Palo Alto, California, USA	33,000	1951–2	3–5

Source: Peterson, J. T. (1971) Climate of the city. In Detwyler, T. R. (ed.), *Man's Impact on Climate.* McGraw-Hill, New York.

the wind speed is less than 5–6 metres/second. The critical wind velocities for the elimination of the heat island effect in some cities are shown in Table 13.4. The urban heat island phenomenon is likely to be less developed in cities in the low latitudes for the following reasons.

1 Because of the prevailing high temperature there is no need for space heating.
2 Most cities in the tropics are not as industrialized or motorized as those in the temperate region. The level of air pollution can be expected to be lower. This implies that more of the terrestrial radiation will escape into space.
3 Urban surfaces in the tropics are less paved and the storm water drainage less efficient than those in the more developed temperate region. This means that less energy will be stored or radiated by the urban surfaces in the tropics. Similarly, more energy will be utilized for evaporation and transpiration with consequently less energy available for warming the air.
4 The urban heat island size and depth are determined by the quantity of heat as well as by city size. Since sizes of cities in the tropics are generally smaller than those of the temperate region, the heat island phenomenon is likely to be less developed in the tropics.

Some consequences of the urban heat island phenomenon are desirable, others are not so desirable. One major desirable effect of the heat island phenomenon is that the need for space heating is less than in the cooler rural environments. This means lower heating bills for urban dwellers. Also, there is a decrease in the frequency and intensity of inversion conditions as a result of increased thermal and mechanical convection over cities. The heat island phenomenon, however, has an undesirable effect in tropical cities which are already warm or even hot so that there is greater need for air-conditioning. Even in temperate cities there is increased demand on summer air-conditioning as a result of the heat island effect. The heat island effect also tends to speed up the process of chemical weathering of building materials. On the other hand, urban warmth is responsible for the earlier budding and blooming of flowers and trees in the cities of temperate regions as well as the longer growing season compared to their rural surroundings (Oke, 1978).

Another aspect of urban climate that has been intensively studied is the pollution of the urban atmosphere by processes of urbanization and industrialization. As in the case of the urban heat island, air pollution has a number of biological, economic, and meteorological implications. Again the vast majority of studies of air pollution have been carried out in cities in the middle latitudes.

Air pollution is the introduction into the air of any substances different from any of its natural constituents. Pollutants may be derived from natural sources or man-made. Naturally occurring pollutants include pollens, bacteria, spores, dust particles derived from volcanic eruptions or raised from the ground by wind as well as smoke particles derived from lightning-caused bush fires. Man-made pollutants include smoke particles, SO_2, CO, CO_2, O_3, hydrocarbons, and various oxides of nitrogen. These pollutants are derived from four principal sources, namely fuel combustion, industrial processes, road traffic, and burning of wastes. Fuels of various types are used for space heating, cooking, lighting, and power generation. Such fuels may be solid (e.g. wood, lignite, coal, and coke), liquid (e.g. petroleum and kerosine), or gaseous (natural gas). Combustion of these fuels gives rise to pollution particularly if combustion is inefficient or incomplete. Most pollutants from fuel combustion consist of smoke particles and SO_2. In many industrial processes substances like chlorine, hydrogen sulphide, flourides, hydrochloric acid, and beryllium are emitted in addition to the normal products of fuel combustion. The worst offenders are the petrochemical industries and cement works. Many of the substances also have an offensive odour. Road traffic is another major source of air pollution in cities. The exhausts from diesel engines contain various oxides of nitrogen while petrol engines emit various mixtures of hydrocarbons. In addition black smoke is emitted by vehicles whose engines are badly maintained or overloaded. The incineration of wastes is the fourth major source of air pollution. The fumes from incinerators contain hydrocarbons, aldehydes, and other organic compounds. Moreover, the fumes are usually too cool to rise far above the ground and are therefore dangerously confined to the air layers close to the ground surface. The type and source of air pollutants in the United States for which data are available are shown in Table 13.5. Carbon monoxide (CO) is the most important pollutant by weight (47%), the least important being oxides of nitrogen (10%). The most important source of pollution in the United States is transportation, contributing 42%. The least important source is the incineration of wastes (5%). Comparable data for countries in the tropics are not available. Transportation can, however, be expected to top the list of pollution sources since the levels of industrialization in these countries are much lower than in the advanced countries of the middle latitudes.

Studies that have been carried out mostly in the temperate countries indicate that the intensity of air pollution in a given area is function of two variables. The first is the rate at which pollutants are emitted while the second is the rate

Table 13.5 Type and source of air pollutants in the USA in 1968

Type of pollutants	% of weight
Carbon monoxide	47
Sulphur oxides	15
Hydrocarbons	15
Particulates	13
Nitrogen oxides	10

Sources of pollutants	% contribution
Transportation	42
Fuel combustion in stationary sources	21
Industrial processes	14
Forest fires	8
Solid wastes disposal	5
Miscellaneous	10

at which these pollutants are dispersed and diluted within the atmosphere. It is the balance between the sources from which pollutants arise and the factors favourable to their dilution and dispersal within the atmosphere that determines whether or not pollutants will constitute a hazard to human health and welfare (WMO, 1965). The rate of emission of pollutants depends on the type and number of sources of pollutants which are in turn a reflection of the size and land use patterns of an urban area. The size and functions of a city provide a rough indication of the number of domestic and factory chimneys as well as the volume of vehicular traffic. On the other hand, the rate of dilution and dispersal of pollutants within the atmosphere depends not only on weather factors but also on other factors such as the height of the chimney stacks, the temperature and speed at which pollutants are emitted, and the topography of the area in which the city is located.

The buoyancy of gaseous pollutants increases as the temperature at which they are emitted increases. Hot gaseous pollutants which are emitted at high velocities from tall chimney stacks can be expected under normal weather conditions to rise and be rapidly diluted so that they would never reach the ground in harmful concentrations (WMO, 1965). Topography is an important factor controlling the rate of dispersal and dilution of pollutants within the atmosphere. Valleys or basins sandwiched amongst hills tend to have a high frequency of inversion conditions which discourage the dispersal of pollutants. On clear calm nights following excessive terrestrial radiation to space from the hillslopes the air becomes cold and dense and drains downslope under gravity, pushing up the warmer air in the valley bottoms and thus creating a temperature inversion condition. Several meteorological factors influence the rate of dispersal and dilution of pollutants within the atmosphere but the most important are wind direction and speed and the vertical temperature profile which determines the stability or otherwise of the air. Generally speaking, high wind velocities

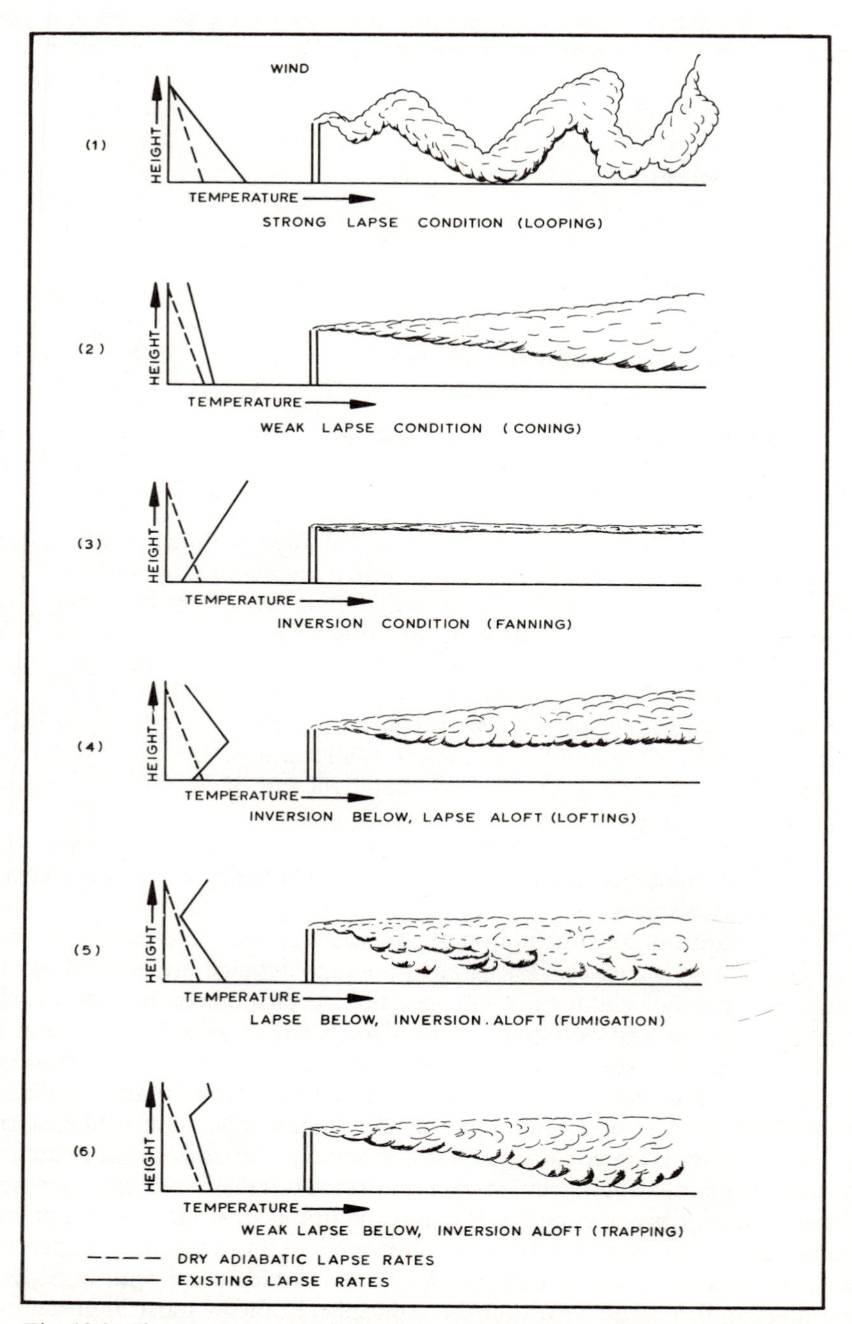

Fig. 13.2. Six types of plume behaviour under various conditions of stability and instability (from Sellers, 1965)

encourage the rapid dilution of pollutants. The greater the wind speed the faster the rate of dilution as a result of the increased turbulence that would be created. Pollutants also tend to concentrate more on the leeward side of cities than on the windward side. Studies of pollution drift indicate that pollutants are often advected from one urban area to another downwind or even from one country to another. The increased acidity of rainfall over Sweden in recent years has, for example, been attribution to the effect of SO_2 and other pollutants advected from the industrial areas of Britain by westerly winds.

The most important meteorological control of air pollution is, however, the stability or otherwise of the air as determined by the nature of the vertical temperature profile. This factor largely determines the behaviour of the pollution plume (see Fig. 13.2). The six geometrical forms of plumes shown in Fig. 13.2 are a function of the vertical temperature and wind profiles. They are as follows:

1 *looping*: this occurs under strong lapse conditions with moderate to high wind velocities and large vertical convective mixing of air. If stacks are low looping can produce high surface concentrations of pollutants.
2 *conning*: this occurs under weak lapse conditions.
3 *fanning*: this occurs under temperature inversion conditions resulting particularly from terrestrial radiation on clear, calm nights.
4 *lofting*: this occurs under conditions of shallow radiation inversion near the ground surface and slight unstable conditions aloft.
5 *trapping*: this is associated with a subsidence inversion situation in which there is a weak lapse near the ground but inversion aloft. Trapping is regarded as one of the worst pollution situations.
6 *fumigation*: this commonly occurs after sunrise when solar heating produces an unstable layer which mixes surface air with pollutants accumulated below night inversion.

Air pollution affects the weather and climate of urban areas in several ways. It plays an important role in the energy balance of urban areas. Pollutants reflect, scatter, and absorb solar radiation. Insolation over cities is 15–20% less than in the surrounding rural areas. Ultraviolet radiation is 5% less in summer and 30% less in winter over urban areas. Sunshine duration is also reduced by 5–15% in urban areas partly as a result of pollution. Many of the pollutants serve as condensation nuclei. There is therefore an abundance of condensation nuclei in the air over cities. Moisture is also amply supplied by cities through evaporation and industrial processes and automobiles that emit large amounts of water vapour into the city air. There is consequently an increase in cloudiness over urban areas as well as in the frequency and intensity of fog. The smoke particles together with fog help to reduce visibility over urban areas.

The observed tendency for rainfall to increase over urban areas is partly due to the pollutants that ensure an abundance of condensation nuclei. Other factors that help to explain the incidence of higher rainfall over urban areas include the addition of water vapour to the city air from various combustion processes, thermal convection over the city heat island, and mechanical turbulence created by the frictional effect of city structures on air flow.

Air pollution has a wide range of mostly adverse effects on plants, man, and his properties. Studies have shown that fruit trees in polluted areas are 10% smaller and produce 10% less fruit with diminished vitamin C content compared to trees in clean air (Maunder, 1970). Pollutants interfere with photosynthesis in plants and affect the energy balance of leaf surfaces. Trees, however, help to filter the air of particulate pollutants but to their own detriment. Carbon dioxide is also withdrawn from the air for photosynthesis and oxygen released as a byproduct to refresh the city air. Trees and parks are therefore desirable in urban areas not only to beautify them but to improve the quality of the air. Trees also help to moderate the urban heat island effect since more energy will be utilized for transpiration and less will be available to heat the air.

Buildings suffer from corrosion as a result of air pollution. Metals are oxidized and there is deterioration of paints. In the United States alone, property damage due to air pollution is estimated to be more than $10 billion a year while damage to crops amounts to over $500 million a year (Maunder, 1970). Pollution reduces visibility with adverse effects on transportation systems. Poor visibility may cause accidents, delays, or cancellations of journeys which may result in financial losses, passenger frustrations, and loss of lives in some cases. It is estimated that more than $66 million is lost every year in the United States as a result of delays and diversion of airline traffic due to poor visibility.

There is evidence that air pollution is an important factor in certain respiratory and lung diseases so common in industrialized cities of temperate latitudes. Table 13.6 shows excess deaths ascribed to the effects of pollution in some cities. Virtually all the excess deaths were due to respiratory and lung diseases particularly chronic bronchitis and its complications. Studies indicate that bronchitis–emphysema is considerably more common among city dwellers than among rural people. In fact, it has been shown that in Great Britain, the larger the city the higher the incidence of bronchitis–emphysema (McDermott,

Table 13.6 Major air pollution episodes and the excess deaths associated with them (after Bach, 1972)

Date	Place	Excess deaths
February 1880	London, England	1000
December 1930	Meuse Valley, Belgium	63
October 1948	Donora, Penn, USA	20
November 1950	Pica Rica, Mexico	22
December 1952	London, England	4000
November 1953	New York, USA	250
January 1956	London, England	1000
December 1957	London, England	700–800
December 1962	London, England	700
January/February 1963	New York, USA	200–400
November 1966	New York, USA	168

1961). Other diseases which have been associated with air pollution include influenza, lung cancer, asthma, and pulmonary heart disease. More medical research is, however, needed to establish fully the link between these diseases and air pollution. It is recognized that air pollution probably only helps to exacerbate rather than cause these diseases. Although not all diseases induced by air pollution are necessarily fatal, the health costs inflicted by air pollution are considerable. They include loss of productivity and working potential as a result of ill-health as well as welfare compensations to workers.

Because of the many undesirable effects of air pollution and the hazard it poses to human health and efficiency, attempts are being made in many countries to control this hazard and enhance the liveability of urban areas. The control of air pollution requires a multidisciplinary approach. Meteorologists, medical doctors, engineers, urban planners, and legal experts all have roles to play. The basic objective of any air pollution control scheme is to maintain the air as clean as possible and not dirtier than a specified quality. It is the job of public health experts to specify the acceptable upper limits of the concentration of various pollutants after necessary studies. The legal authorities can then draw up enforceable codes to make industries and individuals comply with such limits. The various allowable limits for various pollutants in the United States for which information is available are shown in Table 13.7. Both concentration by weight and duration of exposure are important considerations in assessing the hazardousness of pollutants. Engineers can help in the control of air

Table 13.7 Air pollution standards—limits of concentrations permitted in the USA

Pollutant	Weight of pollutants per cubic metre of air
Carbon monoxide	10 mg maximum 8 h concentration
Sulphur oxides	80 μg annual mean 365 μg maximum 24 h concentration
Hydrocarbon compounds	125 μg maximum 3 h concentration
Nitrogen oxides	100 μg mean annual 250 μg maximum 24 h concentration
Petrochemical oxidants (e.g. ozone)	125 μg maximum 1 h concentration
Particulates	75 μg annual mean 260 μg maximum 24 h concentration

Source: After Environmental Protection Agency (EPA) of USA. In Strahler, A. N. and Strahler, A. H. (1974) *Introduction to Environmental Sciences*. Hamilton Publishing Company, USA.

pollution by the design of more efficient combustion processes, dust collectors, and chemical processes to remove or treat pollutants at source as well as other antipollution devices.

Meteorologists can help in controlling air pollution by conducting more research into the manner in which contaminants are transported and diffused within the atmosphere under various stability conditions. Accurate forecasts of weather situations that favour the concentration of pollutants in the atmosphere are of immense help in controlling some of the undesirable effects of air pollution. Studies of the temporal and spatial patterns of air pollution over urban areas are also required in formulating policy measures to alleviate the effects of pollution. Such studies should involve, among others, a comprehensive inventory of emissions to identify major pollution sources and their pollutants and to delineate polluted from unpolluted areas. Meteorologists can, on the basis of their knowledge of controls of air pollution, advise on the judicious location of industrial plants and other major sources of pollution within given urban areas in the light of the prevailing climate. Working hand in hand with meteorologists, urban planners can contribute to air pollution control through proper urban design, involving zoning, proper land use planning and site selection as well as the use of green areas and buffer zones (Bach, 1972).

The hazard posed by pollution to life and property in cities of temperate countries has attracted government attention in the form of various legislative measures designed to control the emission of pollutants. Antipollution laws have been enacted in the United Kingdom, United States, and other countries. There are, however, problems in enforcing such laws. Threshold values of pollutants must be established and pollution levels continuously monitored. There is therefore a need for effective inspection and policing to ensure compliance with regulations as well as public support to ensure the success of any antipollution law. Pollution may also be controlled through the policy of subsidization or that of putting a price tag on pollution. In the first case, industries may be given grants, low interest loans, or tax reliefs by the government to enable them to purchase and operate suitable pollution control equipment. In the second case, industries and private individuals are charged penalty fees in direct proportion to the amount of pollutants emitted into the air or water courses. These will of course have to be monitored and measured at some cost which may render the whole idea economically unsound.

Man and the future of world climate

It has been shown in the preceding sections that man may influence weather and climate deliberately or inadvertently on a local or mesoscale. It is now suggested that man is in fact capable of influencing global climate inadvertently through various processes such as urbanization and industrialization, farming, lumbering, overgrazing, irrigation, drainage of swamps, and the creation of artificial lakes by damming rivers. It has also been suggested that global climate may even be influenced by man deliberately through engineering and other works such as altering the courses of some ocean currents by damming straits, creating vast inland seas in Siberia and Central Africa by damming big rivers, artificially melting the ice in Antarctica and Greenland to reduce albedo. These are all measures that can theoretically alter the global climate but the actual direction of alteration as well as the consequences for mankind are not known and can only be speculated upon. Under these circumstances it is not in the interests of man to seek to deliberately alter global climate through such measures as listed above. Besides, these measures would be very costly to execute and would task man's engineering skill and ingenuity.

Man is, however, right now unwillingly influencing global climate by inadvertently changing the earth's albedo and atmospheric composition (e.g. through urbanization and industrialization) and by artificial heat production. According to the SMIC report (1971) the most widespread modification of the climate in the past has been achieved inadvertently by the conversion of natural vegetation into arable land and pastures. During the past 8000 years, 11% of the land area of the earth has been converted to arable land and 31% of forest land is no longer in its natural state. The modification of the natural vegetation has affected several important climatic parameters such as albedo, the Bowen ratio, and surface roughness as well as the hydrological properties of the surface (Lockwood, 1979). These alterations have implications for the energy and water balances of the earth and the general circulation of the atmosphere on which global climate depends. The various processes that cause a decrease and increase in albedo and Bowen ratio are listed in Table 13.8. Table 13.9 shows the changes in the heat budget after conversion from forest to agricultural use. Computed global average surface albedo values today and 6000 years ago are shown in Table 13.10. All things being equal, an increase in albedo will cause a drop in global surface temperature while a decrease in albedo will raise it.

Table 13.8 Processes that cause increases and decreases in albedo and the Bowen ratio

Albedo	Bowen ratio
Increases	
Desertification	Desertification
Overgrazing in semi-arid regions	Clearing of forests
Burning of grassland in semi-arid regions (slight)	Drainage of swamps
Ploughing of fields (slight)	Urban growth in moist climates
Clearing of forests	
Addition of biological films to water surfaces	
Decreases	
Overgrazing in regions with moderte to heavy rainfall	Irrigation
Man-made lakes and irrigation (slight)	Man-made lakes
Construction of towns (slight)	Urban growth in dry climates
Snow removal	
Deposition of particles on snow	

Table 13.9 Changes in heat budget after conversion from forest to agricultural use (after SMIC Report, 1971)

	Albedo	Net radiation (Wm^{-2})	Sensible heat flux (Wm^{-2})	Latent heat flux (Wm^{-2})
Coniferous forest	0.12	60	20	40
Deciduous forest	0.18	53	13	39
Arable land, wet	0.20	50	8	42
Arable land, dry	0.20	50	15	35
Grassland	0.20	50	20	30

Table 13.10 Global average surface albedo now and 6000 years ago (after Munn and Machta, 1979)

Today		6000 years ago
Northern hemisphere	0.157	0.138
Southern hemisphere	0.151	0.143
The world	0.154	0.141

Another important effect of man's activities on climate is exercised through changes in atmospheric composition particularly the CO_2, ozone, and aerosols content of the atmosphere. There is unequivocal evidence that the CO_2 content of the atmosphere has been increasing over the years as a result of man's use of fossil fuels for energy. Between 1870 and 1970 the total quantity of atmospheric CO_2 is estimated to have increased by 11% (from 294 to 321 ppm) due to the burning of fossil fuels by man. Were it not for the removal of CO_2 from the air by the land biosphere and the oceans, the increase would have been about 20% (Barry and Chorley, 1976). Because CO_2 absorbs and re-emits radiation from the earth and the atmosphere it plays an important role in global temperature. In the absence of other factors, an increase in CO_2 will raise global temperature. Calculations indicate that atmospheric CO_2 may increase to 370 ppm by 2000 AD with a consequent increase in global temperature of 0.5 °C (in the absence of other factors).

Man-made contributions to particulate matter in the atmosphere is estimated to be 30% of the total and this figure could double by 2000 AD (Barry and Chorley, 1976). The overall effect of particulate matter on the lower atmosphere is one of cooling so that the effect is opposite to that of CO_2. Global temperature increased by about 0.6 °C between 1880s and the early 1940s but this was followed by a net cooling of 0.2–0.3 °C by 1970 most probably as a result of increase in man-made particulate matter.

Ozone is another important atmospheric constituent whose concentration in the atmosphere is influenced by human activities. Ozone is found concentrated in the atmosphere mostly between 15 and 35 km above the earth's surface.

Ozone is formed mainly between 30 and 60 km above the earth's surface of collisions between O and O_2 (see Chapter 2). Variations in ozone concentration may occur naturally as a result of fluctuations in solar ultraviolet radiation required to break up the oxygen molecules that will eventually combine with O_2 to form ozone.

Ozone absorbs ultraviolet radiation which is harmful to living things including man. The absorption of dangerous ultraviolet and other short-wave radiation accounts for the warming of the atmosphere at around 50 km. Ozone is continuously being formed and destroyed in the atmosphere. Ozone is destroyed after formation by breakdown into normal oxygen and a single atom or by reactions with pollutants such as the exhaust gases of rockets and high-flying aircraft, or waste from nuclear explosions.

Apart from the formation of ozone in the stratosphere, ozone may be formed a few hundred metres above industrialized cities. Here intense sunshine promotes complex ozone forming chemical reactions involving car exhaust pollutants. Concentrations of ozone greater than 50 parts per hundred million (0.000 05%) by volume near the ground are dangerous. Ozone damages lung tissues and causes irritation of the eyes, nose, and throat. Ozone is also harmful to plants. It makes them susceptible to fungal disease and insect attack. Man should therefore be interested in the quantity and pattern of distribution of ozone in the atmosphere.

Another mechanism by which man may inadvertently influence global climate is through the artifical production of heat (anthropogenic heat). Although averaged over the globe the heat generated by human activities is a trivial fraction of the net solar radiation at the earth's surface; anthropogenic heat can sometimes be of the same order of magnitude as, or even exceed, the net radiation locally or even regionally (Munn and Machta, 1979). A comparison of the magnitudes of several energy sources is given in Table 13.11. Table 13.12 shows that anthropogenic heat in some cities of high latitudes is equivalent to or greater than the local natural radiation receipt. Industrial cities with a compact layout, such as New York City and Moscow, are excessive heat islands. Yet the energy consumption in these and other cities continues to increase.

Table 13.11 Sources and magnitude of some energy releases (after Munn and Machta, 1979)

Energy source	Magnitude (W/m^2)
Solar radiation at outer edge of the atmosphere	350
Net solar radiation at the earth's surface	160
1970 energy production distributed evenly over the globe	0.016
1970 energy production distributed evenly over the continents	0.054
Annual continental net photosynthetic energy	0.016
Annual global energy flow from the earth's interior	0.06
Heat from major cities in the USA in summer	20–40
Heat from major cities in the USA in winter	70–210

Table 13.12 Comparison of anthropogenic heat production in some cities and regions (mainly after Lockwood, 1979)

Location	Area (km^2)	Energy consumption density (Wm^{-2})	Average net radiation (Wm^{-2})	% of energy consumption density of net radiation
Globe	510,000,000	0.016	80–100	0.02–0.016
Continents	135,781,867	0.046	80–100	0.05–0.004
USA (minus Alaska)	7,827,000	0.26	50–110	0.5–0.2
Federal Republic of Germany	246,000	1.36	50	2.7
Los Angeles City, USA	3500	21	108	19.5
Cincinnati City, USA	200	26.2	99	26.5
Moscow, USSR	878	127	42	302
West Berlin	234	21.3	57	37.4
New York City (Manhattan)	59	630	93	677
Sheffield, England	48	19.3	56	34.5
Fairbanks, Alaska	37	18.5	18	100
Hamburg, W. Germany	747	12.6	55	22.9

Several authors have speculated on the possible effect of continued anthropogenic heat releases on the global climate. Some climatic models predict a temperature change of 2–3 °C for a 1% change in heat input. Model results also indicate that changes are twice as large near the poles. It has been calculated that the global solar heat absorption is of the order of 150 Wm^{-2}. Current estimate of the upper limit of anthropogenic heat world-wide is about 1.57 Wm^{-2} whereas global anthropogenic heat emission in 1970 was 8TW. The total additional heat from human activities could thus lead to an average temperature increase of 2–3 °C for the globe and perhaps 10 °C for the polar regions (Lockwood, 1979). Such temperature increases would lead to the melting of polar glaciers and snow fields and a rise in sea level. Coastal areas would be inundated and the global albedo and radiation balance would be altered with concomitant effect on the general circulation of the atmosphere and global climate. Anthropogenic heat sources do not at present to the best of our knowledge influence global climate but they have observable effects on local climates in cities as mentioned earlier.

Conclusion

The atmospheric environment influences man and his activities while man may through his various actions deliberately or in advertently influence weather and climate. Climate must be recognized both as a resource and as a hazard. Climatic management therefore involves the rational use of climatic resources and the avoidance, prevention or minimization of climatic hazards. Sound climatic management is impossible without good knowledge of atmospheric characteristics and processes and their relationships or interactions with biological and man's socioeconomic activities.

Modern man need not like his forebears live at the mercy of weather. But speculative schemes to control regional or global weather and climate should be approached with caution. Not only are they costly but they would task man's engineering skill and many would even doubt man's technological capability to interfere with the atmosphere on such a scale. Even more important is the fact that we know little at present about the likely climatic consequences of such schemes since our present knowledge of the atmosphere and its workings is still far from perfect. In the words of Chandler (1970):

> 'before we can so rashly tamper with climate, we must observe and theorise in order to understand the exact weather science, we must understand before we can predict with a high degree of accuracy before with discretion and sensitivity, we can safely experiment with the atmosphere.'

There is therefore a great need for more observational data and research on our atmospheric environment to facilitate effective and wise management of this important component of our natural environment. In many parts of the world there is still a dearth of weather observations. Examples include oceanic areas, sparsely populated or inaccessible areas (e.g. deserts, mountainous and polar areas) as well as most parts of the tropics. Finance and manpower are clearly problems in many developing countries of the tropics. The World Meteorological Organization and the developed countries of Europe and North America are well placed to help in the provision of necessary equipment and the training of the requisite manpower. However, there is a need to change the laissez-faire attitude of many developing countries to scientific study of weather and climate. This attitude has arisen out of ignorance of the economic and social value of knowledge of weather and climate. Although these countries do not experience changeable weather like the temperate region they suffer from climatic extremes like droughts and floods. In addition, most of the increases in world population are taking place in these countries and their economies are primarily agricultural. The need to improve agricultural production which depends primarily on weather cannot therefore be overemphasized. Increased alteration of the earth's surface resulting from agricultural activities as well as urbanization and industrialization also has an effect on regional and local climate. The need for investment in the scientific study of the atmosphere is therefore no less urgent in the developing countries of the tropics. Because the global atmosphere is a system with linkages and interactions between various components and because of the important role of the tropics in the general circulation of the atmosphere which ultimately controls global weather and climate, the development of weather and climate study in the tropics will benefit the whole of mankind. It therefore demands moral and financial support from the advanced countries in the temperate region of the world.

References

Bach, W. (1972). *Atmospheric Pollution*, McGraw-Hill, New York.

Barry, R. G. and Chorley, R. J. (1976). *Atmosphere, Weather and Climate* (3rd edn). Methuen, London.

Burton, I. *et al.* (1978). *Scope Workshop and Climate/Society Interface*. Toronto, Canada.

Chandler, I. J. (1970). *The Management of Climatic Resources*. Inaugural Lecture delivered at University College, London.

Critchfield, H. J. (1974). *General Climatology*. Prentice-Hall Inc, New Jersey.

Griffiths, J. F. (1976). *Climate and the Environment: the atmospheric impact*. Paul Elek, London.

Hobbs, J. E. (1980). *Applied Climatology: A study of Atmospheric Resources*. Damson, Folkestone.

Landsberg, H. E. (1966). *Physical Climatology*. Gray Printing Company, Pennsylvania.

Landsberg. H. E. (1970). Man-made climatic changes *Science*, **170**, 1265–1274.

Lockwood, J. G. (1979). *Causes of Climate*. E. Arnold, London.

Mather, J. R. (1974). *Climatology: Fundamentals and Applications*. McGraw-Hill, New York.

Maunder, W. J. (1970). *The Value of the Weather*. Methuen, London.

Munn, R. E. and Machta, L. (1979). Human activities that affect climate. In *Proceedings World Climate Conference*, W.M.O.

McDermott, W. (1961). Air pollution and public health. *Scientific American*, **612**, 49–57.

Oke, J. R. (1978). *Boundary Layer Climates*. Methuen, London.

Olgyay, V. (1963). *Design with Climate*. Princeton University Press, New Jersey.

Sellers, W. D. (1965). *Physical Climatology*. University of Chicago Press, Chicago.

SMIC (1971). *Inadvertent Climate Modification*. M. I. T., Cambridge, Mass, U. S. A.

Smith, K. (1975). *Principles of Applied Climatology*. McGraw-Hill, New York.

WMO (1965). A survey of human biometeorology. *Technical Note No. 65*.

APPENDIX:

Revision Questions

Chapter 1

1 Distinguish between the terms (a) 'weather' and 'climate', (b) 'meteorology' and 'climatology'.
2 Discuss the nature and scope of climatology.
3 Write a critique of the traditional approach to the study of weather and climate.
4 Discuss the contributions of the computer and space technology to recent advances in the study of weather and climate.
5 Discuss recent developments in the study of weather and climate in the low latitudes.

Chapter 2

1 Describe the vertical structure of atmosphere.
2 What is the basis for the division of the vertical structure of the atmosphere?
3 Why is the troposphere described as the weather-making layer of the atmosphere?
4 Describe and explain the spatial and seasonal variations in the composition of the atmosphere.
5 Describe the processes of energy and mass transfer within the atmosphere.

Chapter 3

1 What is meant by the solar constant? Give its average value.
2 State (a) the Stefan–Boltzman Law and (b) Wien's displacement law and discuss their significance in radiation studies.
3 What is meant by the altitude of the sun? What factors determine its value?
4 Discuss the role of (a) ozone, (b) carbon dioxide, (c) water vapour, and (d) particulate matter in the earth's radiation balance.

5 Describe and explain the geographical distribution of (a) insolation and (b) net radiation over the earth's surface.
6 Distinguish between 'Rayleigh scattering' and 'Mie scattering'.
7 What is meant by (a) the 'continentality effect' and (b) the 'green house effect'?
8 Distinguish between the concept of radiation balance and that of the energy budget.
9 What is meant by albedo? What are the factors that determine the albedo of a given surface?
10 Examine the nature and global disposition of solar radiation in the earth–atmosphere system.
11 Discuss the significance of solar radiation in climatological studies.
12 Describe how you would measure or estimate the values of the components of the radiation balance equation over a given location.

Chapter 4

1 Describe and explain geographical variations in mean surface air temperature over the globe.
2 What factors determine continentality effect? Discuss the climatic significance of this phenomenon.
3 Discuss with examples the effects of (a) latitude and (b) continentality on seasonal variations in mean surface air temperature.
4 What it meant by the physiological temperature? How is it measured or estimated?
5 Why are the diurnal variations in temperature larger in the tropics than in extratropical areas?
6 How would you assess the degree of human comfort in a particular climate?

Chapter 5

1 Describe the four major controls on horizontal movement of air near the earth's surface.
2 Describe and explain the general circulation of the atmosphere.
3 Distinguish between 'divergence' and 'convergence' and describe the conditions that favour each of them.
4 Explain the locations of the major centres of low and high pressure of the world in (a) January and (b) July.
5 Discuss the role of the tropics in the general circulation of the atmosphere.
6 What is meant by the zonal index and how is it computed?
7 What circulation patterns are associated with (a) high and (b) low values of the zonal index?
8 What are 'jet streams' and what are their relationships with surface weather and climate?

9 Write an essay on monsoon circulation *or* 'Rossby' waves.
10 Describe the origin and characteristics of *two* types of local winds.

Chapter 6

1 What are the major differences between weather systems in the tropics and in extratropical areas?
2 What is a front? Under what conditions does a front form and decay?
3 What is an air mass? What are the three principal factors that determine air mass weather?
4 What are the characteristics of air mass source regions? Determine the source regions from which air masses are likely to affect your home area in the course of the year.
5 Describe the formation, development, and decay of a frontal depression.
6 Describe *three* types of non-frontal depression and explain their occurrence.
7 Discuss the similarities and dissimilarities between the tropical cyclone and the extratropical cyclone.
8 Describe and explain the sequence of weather types associated with the passage of a depression.
9 Discuss the view that weather changes in the tropics are frequent and complex with quite distinct types of weather.
10 What are hurricanes? What are the conditions favourable to their development?
11 Write an essay on thunderstorms.
12 Describe the formation and characteristics of (a) the easterly waves and (b) the West African disturbance lines.

Chapter 7

1 Discuss the significance of atmospheric moisture in the study of weather and climate.
2 Distinguish between (a) evaporation and evapotranspiration and (b) potential evapotranspiration and actual evapotranspiration.
3 What is meant by 'oasis effect'?
4 Discuss the major meteorological factors that affect the rate of evaporation over a given area.
5 Discuss the procedure and problems of measuring evaporation.
6 Give a critical appraisal of two evaporation formulae.
7 Describe the various ways of specifying the water content of a volume of air.
8 Describe the cooling processes which may result in the formation of (a) cloud, (b) dew, and (c) fog.
9 Describe the mode of formation and characteristics of the major cloud types.
10 Discuss the two main theories of raindrop formation.

Chapter 8

1 Discuss the problems of rainfall measurement.
2 Describe the mode of origin and characteristics of (a) convective precipitation, (b) cyclonic precipitation, and (c) orographic precipitation.
3 Explain the location of the major desert areas of the world.
4 In what major ways are the characteristics of precipitation in the tropics different from those in the middle latitudes?
5 Describe carefully the processes that give rise to precipitation.

Chapter 9

1 Describe (a) the radiosonde and (b) radar and explain their usefulness in studies of weather and climate.
2 Carefully describe the stages involved in the production of a surface synoptic weather map.
3 Explain how the tephigram is used in studies of the stability conditions in the atmosphere.
4 Describe the stages in synoptic weather forecasting.
5 Discuss the principles and problems of numerical weather forecasting.
6 Write an essay on problems of synoptic weather forecasting in the tropics.
7 Describe *two* methods of long-range weather forecasting.
8 Discuss the concept of thermal winds and its usefulness.
9 Discuss the usefulness of weather satellites in weather forecasting.
10 Describe *two* methods of assessing the success of weather forecasts.

Chapter 10

1 Discuss with examples methods of assessing past climates.
2 Critically examine the major theories of climatic change.
3 'Climatic variations occur on various time scales.' Elaborate.
4 Consider recent variations in global climate and possible future trends.
5 Briefly trace the history of global climate during various geological periods.
6 Describe the variations in the climate of the African tropics in the last 20,000 years.

Chapter 11

1 Discuss the purpose and problems of climatic classification.
2 Distinguish between genetic and generic approaches to climatic classification and give *two* examples of each.
3 Discuss the principles, strengths, and weaknesses of (a) Koppen's method and (b) Thornthwaite's method of classifying climates.
4 Describe and account for the distinguishing features of tropical and temperate climates.

5 Discuss the relative strengths and weaknesses of *two* approaches to the study of regional climates.

Chapter 12

1 Using suitable examples, discuss the role of climate in agriculture.
2 Discuss the definition of drought. Identify the regions of the world which are most susceptible to drought and suggest suitable methods for combating the problem of drought.
3 Describe methods of artificially altering the microclimate of a crop to improve yield.
4 Examine the influence of climatic conditions on animal husbandry.
5 Discuss the principles and applications of water balance studies to agricultural planning and development.
6 Consider the climatic aspects of crop pests and diseases prevalent in your home country.

Chapter 13

1 Examine the role of climate in human morbidity and mortality.
2 Discuss the influence of climate on native building design and architecture in different climatic zones of the world.
3 What are the main determinants of urban heat island phenomena?
4 Discuss the meteorological considerations which could mitigate some of the undesirable features of urban climate if taken into account in city planning.
5 Examine the meteorological factors which control the incidence and intensity of air pollution in urban areas.
6 Discuss the problems and prospects of purposeful weather modification.
7 Discuss the concept of climate as a resource.
8 With reference to your home area assess the effects of weather conditions on commerce and industry.

Subject Index